Chemistry in the Laboratory

161

Fifth Edition

Paul W. W. Hunter

Amy M. Pollock

&

Sheldon L. Knoespel

Michigan State University

Chemistry in the Laboratory 161

Fifth Edition

Okemos Press
Scientific and Technical Books

Okemos, MI 48805-0085

ISBN 09630471-0-8   978-0-9630471-0-6

Second printing June 2010

Printed by
Keystone Printing Group, Lansing, MI 48906

The logo of the Okemos Press is derived from alchemical symbols for lead ☿ and gold ☉

Contents

Student Information

Name:	
Section number:	
Room number:	
Instructor:	
Day and time of class:	

General Information

Introduction

Objectives

Our knowledge of the properties of chemical systems arises from experiment. Lecture courses are concerned with the results of these experiments as they are systemized in chemical laws and interpreted and explained by chemical theories. In addition to lectures on the laws and theories of chemistry, a thorough study of chemistry requires that you learn and experience some methods of experimental measurement and investigation. It is hoped that you will gain an appreciation of how chemists go about their work in the laboratory and that you will experience some of the excitement of doing experiments in chemistry.

Organization of the Experiments

Each experiment starts with a statement of the objectives for the experiment. This is followed by an introduction that describes the principles involved in the experiment. The next section in each experiment is the procedure to follow in the laboratory. Always read through the experiment before coming to the lab and be prepared and knowledgeable about what you are going to do. Pay attention to the instructions, particularly when hazardous materials are being handled. You will often see margin notes with triangular warning signs emphasizing points at which special care must be taken. Always think about what you are doing and if you have any questions about the procedure, always ask your instructor.

The final section of each experiment is a description of how to go about the calculations that are often necessary in the analysis of your data. Margin notes in this section are intended to assist in the calculations. Values of various physical constants and chemical formulas and relationships are also noted in the margin.

Data Collection

A data sheet for recording data is provided at the end of each experiment. Adhesive labels will be supplied for you to stick on each data sheet to be handed in. The sheet should be removed and handed in to your instructor at the end of each day's experiment. Most experiments in this manual are intended to be done with a lab partner; you should however complete your own data sheet and make a note of your partner's name on your data sheet. Some experiments require that you treat your data graphically. For these experiments you will find graph paper at the end of the experiment. General guidelines on how to draw graphs are described in the next section.

Your experimental results can only be as good as the data you collect. Experimental readings should be done as accurately and precisely as is possible. For example, volumes determined by reading the level of liquid in a buret can be determined to ±0.02 mL. Masses determined on a typical electronic balance can be determined to ±0.001 g. Temperatures measured on a typical digital lab thermometer can be determined to ±0.1°C. However, every measurement involves some error and it is important to take into account any limitations in the experimental procedure. Do not write the results of your calculations with an unjustified precision.

Precision is a measure of the reproducibility of an experimental measurement. In many of the experiments in this manual you will repeat measurements; the closer they are, the more precise the measurement. Accuracy is a measure of how close the experimental measurement is to the true value.

If the experimental procedure involves some systematic error (for example, if you always make the same mistake or if an instrument is calibrated incorrectly), then the results may be precise but will not be very accurate. On the other hand, if the experimental procedure involves only random errors, then the result will be accurate but may not be very precise. This is why it is important to repeat experiments and examine the data you collect carefully.

Graphing Experimental Data

It is important that you graph experimental data correctly *with neatness and precision*. If the line is a straight line use a ruler! The following is a general procedure:

1. Scientific experiments often consist of measuring what happens to one property of a system as another property is varied. The property that is varied is referred to as the *independent variable* and the property that is measured is the *dependent variable* since it depends upon the first.

    The independent variable is plotted along the horizontal or *x*-axis (the abscissa) and the dependent variable is plotted on the vertical or *y*-axis (the ordinate), as shown in Figure 1.

2. Label the axes and write the units next to the labels (for example, g, mL, mA, sec).

3. Examine the range of values to be plotted on the graph and divide each axis into convenient divisions so that you use as much of the graph paper as possible. The larger the scale the greater the precision. You can use the paper vertically or horizontally and it is not always necessary to include the origin on the graph. Note also that it is not necessary to make one square equal to the same number of units on each axis.

4. Plot your experimental data on the graph paper as precisely as possible.

Figure 1. A plot of $y = mx + b$.

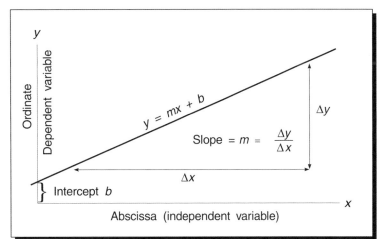

The quantity m is called the gradient or slope of the line.

5. Determine the purpose for drawing the graph. Is it to determine a general relationship between two variables? If so, you will probably be required to draw the best straight line through the points using a ruler. Note that the line may go through all the points, only some of them, or even none of them and the origin may or may not be on the line. The line must be straight and continuous; do not connect the individual points with a zigzag line.

   In some cases a graph may be drawn to determine the area under the curve rather than the overall relation between two variables. In this case the curve will not necessarily be a straight line.

6. You may be required to determine the gradient, or slope, of a straight line on the graph. The gradient is the ratio between the vertical rise (Δy) and the horizontal run (Δx). Pick two points on the straight line as far apart as possible; measure the corresponding rise and run on the two axes and then determine the ratio $\Delta y/\Delta x$.

Laboratory Equipment

Your instructor will describe and explain the equipment available in your laboratory. You will be assigned an equipment locker or drawer. Check the equipment carefully and make sure that all items are present and in good condition. Check the glassware for cracks and chips. You are responsible for this equipment. Some common laboratory glassware is shown at the end of this section. For some experiments additional equipment will be supplied.

Safety

Safety in the laboratory is of paramount importance. For the protection of everyone in the laboratory you must obey common sense safety rules. Your instructor will require you to read a list of safety regulations and sign that you have done so. These rules will include the following and may include others. Laboratory instructors are required to enforce safety regulations without exception.

1. Approved safety goggles must be worn when experiments involving chemicals are in progress in the laboratory. Contact lenses must not be worn in the laboratory—various vapors will accumulate under the lens and cause injury even after leaving the lab.

2. Treat all chemicals with great care. If you get a chemical on your skin, wash it off *immediately* and get medical attention if necessary. Do not look into the open end of a test tube or reaction vessel and do not point a test tube at your neighbor. Some fumes and vapors are poisonous; do not breathe these vapors; perform experiments involving them under the hood. Clean up all chemical spills promptly; get help from your instructor if necessary.

3. Learn the location of emergency equipment both inside and outside the laboratory. This includes the fire extinguishers, fire blankets, overhead showers, hand showers, eye-wash bottles, any other first aid supplies, and emergency telephones. Learn the location of the material safety data sheets (MSDS's).

4. Notify your instructor *immediately* if you have an accident in the laboratory. Consult your instructor *immediately* if you have any questions about safety.

5. You must wear appropriate clothing and shoes in the laboratory. Cotton jeans, for example, afford better protection than synthetic fabrics. Shoes, rather than sandals, protect your feet from spills.

6. Eating, drinking, and smoking are forbidden in the laboratory.

7. Chemical equipment is often fragile and sometimes expensive; treat it with care.

8. Do not sit on lab benches; do not perform unauthorized experiments; and do not interfere with other students' experiments. Do not remove any chemicals or equipment from the laboratory.

9. Used chemicals must be disposed of properly. Use approved waste containers for hazardous waste. Use the proper waste containers for broken glass and paper. Don't mix broken glass with waste paper.

General Information

In case of an accident

- Remain calm and get assistance from your instructor.

- Report all injuries to your instructor. Your instructor will assist you.

- Get treatment for all injuries, cuts, burns, etc. Transportation will be provided if required.

First aid

Emergency treatment should be limited to the prevention of further injury until professional assistance can be obtained. The following accidents require immediate action to prevent further injury:

- **Chemical in the eye:** Hold the eye open and flush it with water *immediately* and continue until professional help arrives.

- **Chemical on the skin:** Flush with large volumes of water from a hand-held or overhead shower *immediately*. Do not be concerned with getting your clothes wet. Do not use any ointment, soap, or chemical reagent.

- **Severe bleeding:** Apply pressure directly to the wound.

Common causes of accidents

Be alert and avoid these accidents:

- Most laboratory accidents are cuts from broken or chipped glassware, glass tubing, or thermometers.
- Chemical burns from concentrated acids are the second most common cause of injury, followed by burns from hot plates or hot liquids.

Additional Information

Additional information for your course this semester will be provided on a separate handout by your instructor. Please read it carefully. This handout will include information about such things as:

- Changing your class schedule.
- Checking out and checking in your equipment.
- What to do if you miss your laboratory class.
- Your prelab discussions or problem sets.
- Where and how to get assistance.
- How your grade will be determined.
- The instructors and the staff for the course.

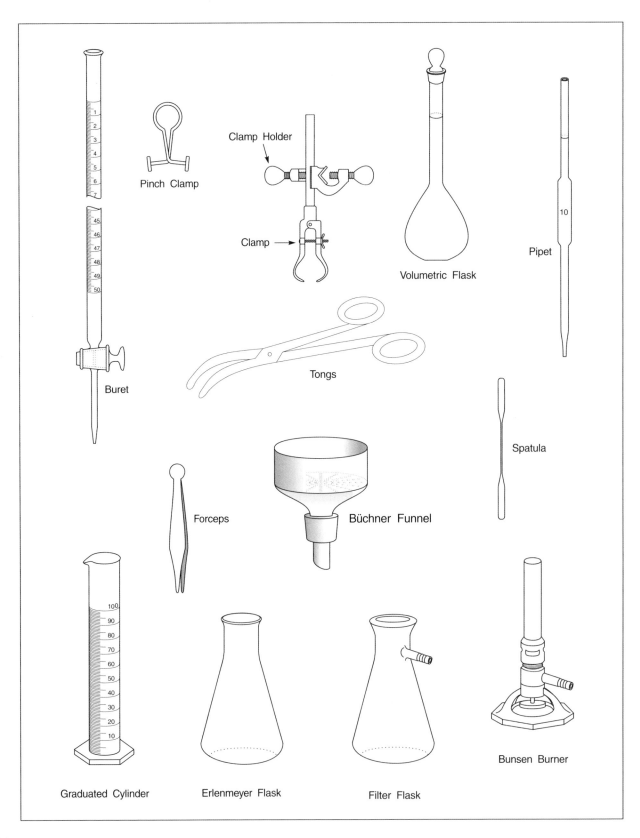

Pinch Clamp

Clamp Holder

Clamp

Volumetric Flask

Pipet

Buret

Tongs

Spatula

Forceps

Büchner Funnel

Graduated Cylinder

Erlenmeyer Flask

Filter Flask

Bunsen Burner

Determination of the Mass, Volume, and Density of Metal Samples

Objectives

In this experiment you will determine the density of a metal. You will weigh samples of the metal and measure the dimensions of the samples to determine their volumes. You will also determine the volumes by displacement of water. By comparing your results with those of other students, you will observe the difference between precision and accuracy. The results of all students in your class will be analyzed by computer and the standard deviation calculated. You will learn how many significant figures you can reasonably use to report your data.

Introduction

$$\text{Density} = \frac{\text{Mass}}{\text{Volume}}$$

Density is a physical property of all substances—gases, liquids, and solids. It relates the mass of a substance to its volume and is expressed as mass per unit volume. A common unit for expressing the density of a solid is grams per cubic centimeter.

Mass and volume are *extensive* properties whereas density is an *intensive* property. An *extensive property* is one that depends upon the quantity of substance in the sample being examined. An *intensive property* is one that is the same no matter how much of the sample is examined.

Density is temperature-dependent (objects tend to expand on heating) so temperature must be reported along with the density. For solids, densities are usually reported at 20°C (approximately room temperature).

The metal blocks used in this experiment are all made of the same metal and, since density is an intensive property, the experimental values for the density determined for all metal blocks and by all students in your class should be the same even though the metal blocks themselves have different shapes and sizes. However, the experimental values will differ slightly because of errors in the different measurements.

Kinds of Errors

There are two kinds of errors: *a systematic error*, caused by a faulty procedure, and a *random error*, caused by a limit to the reproducibility of measurements. For example, a systematic error would result if a balance was incorrectly calibrated; a random error might arise when you estimate fractions of a division on the scale of an instrument. Systematic errors are sometimes difficult to detect and correct. Random errors, on the other hand, are more obvious and can be analyzed statistically.

Two terms often used to describe the exactness of experimental results are *precision* and *accuracy*. **Precision** is a measure of how near a series of experimental results are to one another. In other words it is a measure of the reproducibility of an observation. **Accuracy** is a measure of the correctness of the result—how near the experimental result is to the true value.

Figure 1.

Precision contrasted with accuracy. Making a measurement is somewhat like shooting at a target: the results may cluster close together (high precision), but a systematic error in an instrument may make the results miss the target (low accuracy). Poor use of a good instrument can cause the results to scatter, but the average should be on target. A good measurement, like good target shooting, gives results that cluster closely around the true value.

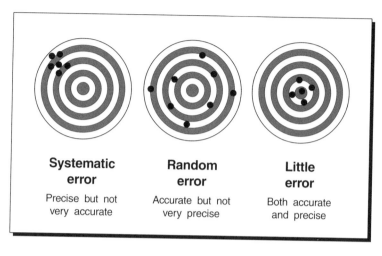

Systematic error	Random error	Little error
Precise but not very accurate	Accurate but not very precise	Both accurate and precise

A good **analytical balance** can weigh to the nearest 0.0001 gram.

A **triple beam balance** usually can measure to only 0.1 or 0.01 gram at best.

Results which are **precise** but **not accurate** indicate some systematic error in the experimental procedure. A series of results which are **accurate** but **not precise** point to random errors in the experimental procedure—see Figure 1.

Experimental results should not be reported with greater precision than the experimental technique warrants. For example, mass determined by an *analytical balance* is known much more precisely than the mass determined by a *triple beam balance*.

Significant Figures

The precision in a number is reflected by the number of *significant figures* written in the number. The more significant figures in a number, the greater the precision of the measurement it represents. For example, the mass of an object weighed on an analytical balance might be written as 3.2719 grams (five significant figures) whereas the mass of the same object weighed on a common laboratory electronic balance can only be written as 3.272 grams (four significant figures). A good laboratory triple-beam balance cannot detect mass differences beyond the first or second decimal place (3.3 or 3.27). If you're not sure how to tell how many significant figures are in a number, study the information in Figure 2.

Figure 2. Significant figures

How many significant figures are in a number?

1. If there is a decimal point in the number, draw an arrow from left to right until you come to the first nonzero digit. All digits you did *not* draw through are significant.

   Example: → 8230.1   5 significant figures
             ← 0.007056   4 significant figures

2. If there is no decimal point in the number, draw an arrow from right to left until you come to the first nonzero digit. All digits you did *not* draw through are significant. You can't tell about the zeros unless the number is written in scientific notation.

   Example: 9132000 ←   4 significant figures
                           (or as many as 7 if zeros are significant)

Some Statistical Concepts

In this experiment you will obtain various values for the density as determined by all students in the class. You will analyze these data to find the standard deviation. Here are some terms used in statistical analysis that you should understand:

Mode: The most frequently observed value in the sample taken.

Median: The central value—there are an equal number of experimental values above and below this value. Therefore to find the median, list the experimental values in ascending or descending order. By counting, find the value at the midpoint.

Mean: The average value. The arithmetic mean, $\overline{X}$, of a set of measurements is the number obtained by dividing the sum of all measurements by the number of measurements in the set.

Deviation: The divergence of a particular experimental value, X_i, from the mean.

Standard deviation: The dispersion or spread of observations around the average in a sample. It equals the square root of the sum of the squares of the deviations divided by the total number of experimental values minus one. Thus:

$$\text{the standard deviation} = s = \sqrt{\sum_{i=1}^{n} \frac{(X_i - \overline{X})^2}{n-1}}$$

where s is the standard deviation
   X_i is the i th experimental value
   n is the number of experimental values in the sample

The standard deviation is a measure of the reproducibility of an experimental result. The smaller the value of the standard deviation, the more precise is the measurement. The final numerical result of an experimental measurement is often written as the mean ± the standard deviation.

In this laboratory class you will initially record your experimental values with more significant figures than is warranted by the precision of your measurements. Then, after you have analyzed the data of everyone in the class, you will have enough information to report the correct number of significant figures for the density of the metal in the sample you measured.

Procedure

You must work on your own for this experiment.

1. Your instructor will give you three blocks of metal with different cross sections—circular, rectangular, and hexagonal. The masses and volumes are unknown.

2. Weigh the metal blocks on a triple beam balance as precisely as possible. Do *not* use an electronic analytical balance. Determine the mass in grams to two decimal places. Repeat the weighing on a different triple beam balance and average the results. Record your data in Table 1 of the data sheet.

3. Examine your metal samples and measure the appropriate dimensions for the different cross sections as precisely as possible using a vernier caliper. Instructions for using the vernier scale are provided at the end of this procedure. Your instructor will assist you if you have difficulty. Be sure to use the scale for exter-

nal (outside) dimensions. Record your results in *centimeters* (*not* millimeters) to two decimal places on the worksheet (Figure 3). Formulas for the cross-section areas of the circular, rectangular, and hexagonal blocks are given in Figure 3.

4. Record the cross-section areas of your three samples in square centimeters with four decimal places in Table 2 on your datasheet. Measure and record the lengths of the metal samples.

5. Calculate the volumes of the metal blocks. The volume equals the cross section area multiplied by the length. Consult your instructor if you are unsure how to proceed. Record the volumes in Table 2 in cubic centimeters with two decimal places.

6. Now calculate the density of the metal for all three samples and record the results in Table 2 with two decimal places.

7. You can check your calculation of the volumes of your metal samples quite easily by measuring the quantities of water they displace. From your drawer select the smallest beaker in which the samples will fit (one at a time) and in which they can each be completely immersed. Fill the beaker with water to a level which you know will cover your samples and which coincides with some convenient reference mark on the beaker. Then pour the water from the beaker into a graduated cylinder to measure its volume *as precisely as possible.* Figure 4 shows how to read the level of liquid in a graduated cylinder. Read the level of liquid at the *bottom* of the meniscus; keep your eye level with the meniscus to avoid a parallax error.

Figure 4. How to read a meniscus

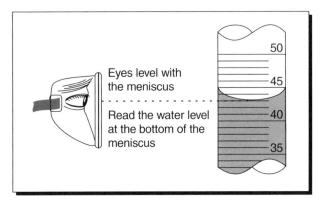

8. Now, one at a time, place the metal samples in the beaker and fill the beaker to the *same level* as before. Each time, the metal objects must be completely immersed. For each metal block, measure the volume of water in the beaker the same way by pouring the water into a graduated cylinder. The difference in the two volumes of water is the volume of your sample. Record your measurements in Table 3 on the data sheet.

9. At this point collaborate with three nearby students—each student should have three sets of data, one for each metal block. Record their names and data from their Tables 1 and 3 in your Table 4. For each sample, plot a point on a graph of mass *vs.* volume—see Figure 5. Label each point *neatly* with the name of the student to whom it belongs. *Use the volume data from Table 3 (water displacement), not Table 2 (dimensions).* Include the origin on your graph and use it as one point. You should have twelve points, plus the origin, on your graph.

You will be graded on how neatly and precisely you draw your graph (40% of your score).

This grade is independent of the accuracy of the result—which is another 40%.

10. Draw the best *straight line* through all points on the graph and determine the slope of this line, which represents density (mass divided by volume). Notice that an accurate result can be obtained from a set of imprecise measurements if the errors are random. Record the result on your data sheet in Table 5.

11. Now return to your individual determination of the density. You now should have an idea what the true density is. If your original measurement of any volume by multiplying dimensions seems incorrect, go back and do it again. Using your data in Table 2, calculate the average density of your samples in grams per cubic centimeter with four decimal places. Record your result in Table 6 on the data sheet. Rewrite the result with the correct number of significant figures. Show your data sheet to your instructor who will initial the data and enter it into the computer. The computer will calculate the mean and the standard deviation for the class. You can record this information in Table 7.

12. Calculate the deviation of your result from the mean value for the class and record it in Tables 6 and 7. Is your result acceptable? Comment on any systematic or random error you noticed among the values in the class. Complete your data sheet, and the graph, and hand it in before leaving the laboratory.

Figure 5. Determination of density (D) from Tables 1 and 3

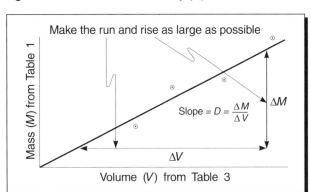

Instructions for the Use of Vernier Calipers

1. Vernier calipers are instruments for making quite precise measurements of the dimensions of relatively small objects. Figure 6 illustrates a simple version of this instrument. Place your sample between the sliding jaws of the calipers and adjust them to a firm fit on the sample.

Figure 6. Vernier calipers

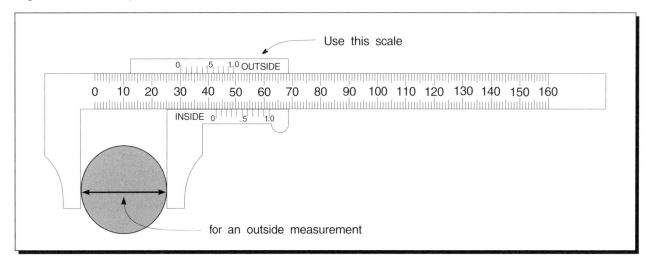

2. A vernier scale is a small auxiliary scale attached to the main scale of an instrument and calibrated to indicate fractional parts of subdivisions of the larger scale. Read the required dimension from the vernier scale of your calipers as illustrated in Figure 7.

    The zero line of the vernier scale (the scale on the sliding part of the caliper) falls between 20 and 21 on the millimeter scale. This means that the required dimension lies between 20 and 21 mm.

    To find the next decimal place (the nearest tenth of a millimeter), compare the vernier scale with the millimeter scale. Only one line on the vernier scale coincides with a line on the millimeter scale. In the example in Figure 7, that line is the sixth one from the zero of the vernier scale. The dimension is therefore 20.6 mm or 2.06 cm.

Figure 7. How to read a vernier scale. This one reads 20.6 mm

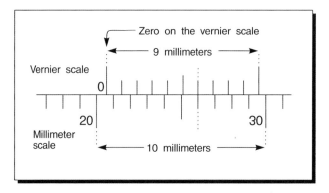

Figure 3. Worksheet for the dimensions and cross-section areas of the metal samples

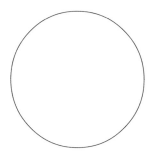

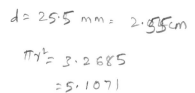

$d = 25.5\,mm = 2.55\,cm$

$\pi r^2 = 3.2685$

$= 5.1071$

Circle: area $= \pi r^2$

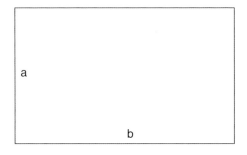

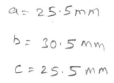

$a = 25.5\,mm$

$b = 30.5\,mm$

$c = 25.5\,mm$

Rectangle: area $= a \times b$

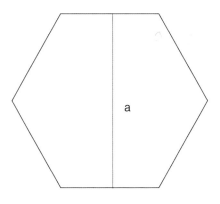

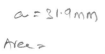

$a = 31.9\,mm$

Area $=$

Hexagon: area $= \frac{1}{2}\,a^2\,\sqrt{3}$

17

Table 7. Results of computer analysis

Mode		g/cm³
Median	2.6900	g/cm³
Mean	2.6887	g/cm³
Standard deviation	.1201	g/cm³
Deviation of your result from the mean		g/cm³
Comments		

The Separation of Recyclable Plastics by Density

Objectives

In this experiment you will determine how to separate recyclable plastics through differences in their densities. You will measure the densities of various liquids and then you will use these liquids to determine the densities of samples of various common plastics.

Introduction

As you learned in Experiment 1, density is an important physical property of gases, liquids, and solids. It relates the mass of a substance to its volume and is expressed as mass per unit volume. A common unit for expressing the density of a solid is grams per cubic centimeter. A unit often used for liquids is grams per milliliter.

$$\text{Density} = \frac{\text{Mass}}{\text{Volume}}$$

Density is an intensive property and can be used to identify substances. An early legendary example of the identification of a substance by density occurred when King Hiero of Syracuse suspected that a crown he had requested made of gold had had some silver substituted for the gold. He asked Archimedes (287–212BC) to find out without destroying the crown. The legend has it that Archimedes discovered how to do this while in his bath and ran naked through the streets of Syracuse crying out "Eureka, Eureka".

Archimedes discovered that a submerged object displaces a volume of liquid equal to the volume of the object. On the other hand, a floating object displaces an amount of liquid equal to the mass of the object. He was able to measure both the mass and the volume of the crown and calculate the density of the material and compare it to the densities of gold and silver.

Archimedes' principle states that an object in a fluid is buoyed up by a force equal to the weight of the fluid displaced. This rule applies to both floating and submerged objects. When the object floats, the mass of liquid displaced equals the mass of the object—the force down is equal to the force up. When the mass of liquid displaced is less than the mass of the object, the object sinks.

The volume of displaced fluid depends on the density of the fluid. For example, someone floating in the Great Salt Lake in Utah will displace less fluid than the same person floating in one of the Great Lakes because the density of salt water is greater than that of fresh water. When an object has exactly the same density as the fluid into which it is placed, it will neither sink nor float but rather hover at any point within the fluid. This is the basis of this experiment.

Density can be used to separate plastics before they are recycled to make new products. This is done by shredding the plastics and dumping them into a very low density solution. The pieces that float are removed and remainder dumped into a slightly more dense material. Again the pieces that float are removed. If this process is continued using solutions of increasing density, it is possible to separate the plastics by their density and therefore by their type.

You will investigate a group of plastics that are commonly identified by recycling codes imprinted somewhere on the container, usually on the bottom. The six most common codes are shown in Table 1, along with the chemical name, the abbreviation used on the product container and the chemical structure of the monomer, or the single repeating molecule, that makes up the polymeric plastic.

Figure 2. Recyclable Plastics: their codes, structures, and sources.

Code	Structure of monomer	Sources
1 PETE	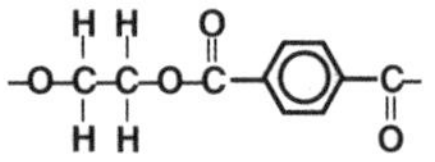 polyethylene terephthalate	Plastic soft drink bottles, mouthwash bottles, peanut butter and salad dressing containers
2 HDPE	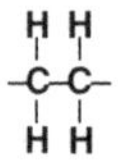 high density polyethylene	Milk, water and juice containers, grocery bags, toys, liquid detergent bottles
3 V	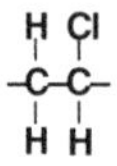 polyvinyl chloride	Clear food packaging, shampoo bottles
4 LDPE	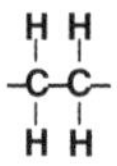 low density polyethylene	Bread bags, frozen food bags, grocery bags
5 PP	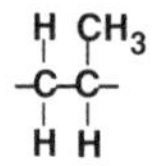 polypropylene	Ketchup bottles, yogurt containers, margarine tubs, and medicine bottles
6 PS	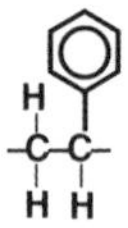 polystyrene	Videocassette cases, CD jewel cases, coffee cups, plastic tableware, and many food take-out containers

Procedure

Work with a partner for this experiment but prepare and submit individual data sheets. Write the name of your partner on your data sheet.

1. Your instructor will assign four different plastics for you to investigate. You might like to check them in the second column in Tables 2 and 3 on your data sheet. The liquids used are water, isopropyl alcohol, saturated sodium chloride solution, and saturated potassium iodide solution. These liquids have different densities.

2. The first step is to determine the densities of three of the four liquids. (You can assume that the density of the distilled water is 1.00 g mL^{-1}.) This requires the

All solutions containing potassium iodide and/or isopropanol must be disposed of properly in the appropriate waste container.

Water and sodium chloride solutions may be disposed of down the sink.

Use small quantities of the liquids! Do not use more than 100 mL of each liquid for the entire experiment!

· weight of graduated
  32.350g cylinder

32.316g

measurement of the mass and the volume of samples of the three liquids. Collect 50 mL samples of the three liquids in three separate beakers.

3. Clean and dry a 10 mL graduated cylinder. Weigh it accurately to determine its mass. Choose one of the liquids and add a small volume to the graduated cylinder. It is recommended that you use a dropping pipet to measure the volume of the liquid as precisely as possible. Measure the mass on an electronic balance. Add some more of the same liquid, measure the volume, and determine the mass again. Repeat the process again with the same liquid. Record these mass and volume data in Table 1 of the data sheet and calculate the average density for the liquid. Return the liquid to your beaker.

4. Clean the graduated cylinder thoroughly and repeat step 3 for the other two liquids. Calculate the average densities of these liquids and summarize the results in Table 1.

5. Now test each of the four samples of plastics you have been assigned in each of the four liquids. You should determine the liquids on which each sample floats and the liquids in which each sample sinks. When you have done this you will know that the density of the plastic sample lies between the density of the liquid on which the sample floats and the density of the liquid through which the sample sinks.

When you test each sample, use a toothpick or piece of wire (such as a straightened paper clip) to push the sample below the surface of the liquid. The sample may stay on top of a liquid simply due to the surface tension of the liquid. Also make sure that the sample has no air bubbles clinging to it. If the sample has a lower density than the liquid, it will return to the surface. If its density is higher, it will continue to sink through the liquid. Record the results in Table 2.

6. Your next step is to determine as accurately as possible the exact density of the plastic sample. To do this you should prepare a liquid mixture in which the plastic sample neither sinks nor floats, but stays at the position at which it is placed in the liquid mixture. To make the mixtur, use the two liquids just above and just below the density of the plastic sample—you determined these two liquids in Step 5.

Mix precise amounts of the two liquids, varying the quantities in a logical way, until you obtain a liquid mixture in which the plastic sample floats wherever it is placed. For example, you could start with 2 mL of the denser liquid and add the less dense liquid drop by drop until the plastic sample neither sinks nor floats. The sample should not drift upward or downward in the liquid. Summarize your results in Table 3 on your data sheet.

7. Now you must determine the density of the liquid mixture that has the same density as your plastic sample. The easiest way to do this is as follows:

Determine the mass of an empty, clean, and dry 10-mL graduated cylinder. Use the same 10 mL graduated cylinder as before—the mass should be the same! Add a precise volume (for example, 5 mL) of your liquid mixture to the graduated cylinder. Again, use a dropping pipet to achieve a precise volume. Now weigh the graduated cylinder and hence determine the mass of the liquid mixture. From the mass and the volume, calculate the density of the liquid mixture. This is the density of the plastic sample. You may wish to repeat this, adding more of the liquid mixture to the graduated cylinder and determining the mass again. Average the densities.

8. Clean the graduated cylinder thoroughly and repeat the process for the other plastic samples you have been assigned.

The Stoichiometry of the Reaction between Baking Soda and Vinegar

3

Objectives

In this experiment you will determine the stoichiometry of a chemical reaction. You will investigate the reaction of baking soda (sodium bicarbonate) with vinegar (acetic acid). The reaction will be monitored by careful measurement of the volume of carbon dioxide released when varying quantities of sodium bicarbonate are treated with varying amounts of acetic acid. This type of analysis is called eudiometry.

Introduction

A chemical equation represents a chemical reaction. Examination of the equation provides considerable information about the reaction, especially the relationship between the quantities of reactants required and the quantities of products formed. This relationship between the reactants and products is called the *stoichiometry* of the reaction.

The word *stoichiometry* is derived from the Greek meaning the "measurement of elements."

There are two main types of chemical reactions—those classified as *acid-base* reactions and those classified as *reduction-oxidation* (redox) reactions. The definitions of the terms *acid* and *base* vary and the distinction between acid-base and redox reactions sometimes is blurred. However, if there is an obvious change in the oxidation numbers then the reaction is classified as redox. If, on the other hand, there is no change in oxidation number, then the reaction is classified as either acid-base or a special type of reaction called *metathesis*.

A *metathesis* reaction is sometimes called a double-displacement reaction.

Sodium bicarbonate is a base, acetic acid is an acid, and the two react to form a salt, sodium acetate. Carbon dioxide and water are also produced. The equation that represents this reaction is:

$$NaHCO_{3(s)} + CH_3CO_2H_{(aq)} \rightarrow NaCH_3CO_{2(aq)} + H_2O_{(l)} + CO_{2(g)}$$

This method of representing a chemical reaction was developed in 1813 by Jöns Jacob Berzelius (1779–1848) from Sweden. Berzelius introduced this logical way of representing compounds and equations to replace the more pictorial and arcane symbolism of Lavoisier and Dalton (see Figure 1 for example). Chemical equations immediately became more understandable.

The coefficients in front of the symbols for the components in a chemical equation indicate the number of moles of reactants required and the number of moles of

Figure 1. Symbols for water

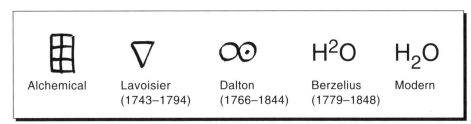

| Alchemical | Lavoisier (1743–1794) | Dalton (1766–1844) | Berzelius (1779–1848) | Modern |

products produced in the reaction. In this equation the coefficients are all equal to one.

According to the equation, one mole of sodium bicarbonate requires one mole of acetic acid, and *vice versa*. You will test this stoichiometric relationship in this experiment.

One *mole* of anything is Avogadro's number of that thing. A mole is simply a counting unit just like a *dozen* (12), *score* (20), *gross* (144) or *ream* (500). The difference is that these units are relatively small whereas a mole represents a *very* large number (6.022×10^{23}). For chemical substances, the mass of one mole equals the unit mass (the atomic, molecular, or formula mass) expressed in grams. For example, there are Avogadro's number of ^{12}C atoms in 12 grams of carbon-12. Twelve grams is the *molar mass* of carbon-12.

Very often, in chemical reactions, the reactants are not present in the correct stoichiometric ratio. There will be too much of one of the reactants and too little of another reactant. The reactant that is used up first will determine how much product (in this case carbon dioxide) will be formed. This reactant is called the *limiting reactant* (because it limits the amount of product formed).

In this experiment, you will systematically vary the amounts of baking soda and vinegar used. In some cases, the baking soda will limit the amount of product formed; in other cases the vinegar will limit the amount of product. By measuring the amount of product you will be able to discover the stoichiometry of the reaction. This analytical procedure is called the *method of continuous variation*.

According to Avogadro's Principle, the volume of a gas is directly proportional to the number of moles at constant temperature and pressure. If the volume of carbon dioxide gas is measured at a known temperature and pressure, the number of moles of the gas can be calculated. The chemical equation written at the beginning of this introduction indicates that one mole of carbon dioxide is released for every mole of sodium bicarbonate, or every mole of acetic acid. This stoichiometric relationship will also be tested in this experiment.

Procedure

Work with a partner for this experiment but prepare and submit individual data sheets. Write the name of your partner on your data sheet.

An *aliquot* is a representative portion of some substance, often in solution.

1. Rinse a two-liter pop bottle with distilled water. Shake out the bottle and then add about 250 mL of distilled water. (This amount need not be precise—use a graduated beaker).

2. Select one of the experiments listed in Table 1 at random and dispense the appropriate amount of glacial acetic acid from the autopipetter into your pop bottle. The autopipetter should be set to deliver 3 mL at a time—do *not* change this setting. Note that all amounts of acetic acid required are multiples of 3 mL. You may have to dispense several 3 mL aliquots. Swirl the bottle to mix the solution.

3. Weigh the corresponding amount of sodium bicarbonate (listed in Table 1) in a beaker. The mass need not be exactly equal to quantity listed but should be reasonably close. Weigh the sample precisely and note the mass.

Table 1. Quantities of acetic acid and sodium bicarbonate used.

Experiment	Volume of Acetic Acid	Mass of Sodium Bicarbonate
1	24 mL	9 g
2	21 mL	13 g
3	18 mL	17.5 g 17.49
4	15 mL	22 g
5	12 mL	26.5 g
6	9 mL	31 g
7	6 mL	35 g

Be careful with glacial acetic acid. If you accidentally get a drop on your skin, wash it off immediately with water.

4. Obtain a new balloon from your instructor and blow it up once to stretch the latex. Use this same balloon for all your experiments.

5. Carefully transfer your sample of sodium bicarbonate to the balloon.

6. Without allowing any sodium bicarbonate to fall into the bottle, attach the balloon to the top of the pop bottle. When the balloon is securely attached, tip the sodium bicarbonate into the pop bottle. Jiggle the balloon to ensure that all the sodium bicarbonate goes into the bottle. The sodium bicarbonate will react with the acetic acid and release carbon dioxide gas.

7. Swirl the bottle to make sure that the reaction is complete. In those experiments with an excess of sodium bicarbonate, some may remain undissolved.

8. When the reaction has finished, hold the balloon upright and carefully measure its circumference using a piece of string. Both you and your partner should measure the circumference *independently*; then, if you agree fairly closely, average your two measurements. If you do not agree closely, you should *both* measure the balloon again. Note your measurements of the circumference in centimeters with 3 significant figures in Table 2 on your data sheet.

A meter rule is provided to measure the length of string.

If you use inches, note that 1 inch equals 2.54 cm exactly.

9. Remove the balloon and rinse out the pop bottle. It is important to remove all the acid from the bottle. Do not get the inside of your balloon wet! Go back to Step 1 and repeat the experiment using a different amount of sodium bicarbonate and acetic acid. You will achieve better results if you do the experiments in random order, for example, in the order 1, 7, 3, 6, 2, 5, 4 rather than the sequence 1, 2, 3, 4, 5, 6, 7.

The experiments are done in random order to average any effects of the balloon stretching.

Plotting the Results

1. Before plotting the graph of your experimental results, some calculations are necessary. Each mole (Avogadro's number of molecules) of acetic acid has a volume of 57.3 mL. So the volume of acetic acid used is divided by 57.3 to obtain the number of moles used. This has already been done for you in Table 2 on your data sheet.

2. The next calculation is to divide the mass of sodium bicarbonate you used by its molar mass (84.01 g/mol) to determine the number of moles. Enter the results in Table 2.

The mole fraction of component A (χ_A) in a binary mixture of A and B equals the number of moles of A (n_A) divided by the total number of moles ($n_A + n_B$).

3. The third calculation is to determine the mole fraction of the sodium bicarbonate; in other words the fraction of the total number of moles of reactants that is sodium bicarbonate. Add the number of moles of acetic acid to the number of moles of sodium bicarbonate to obtain the total number of moles and then divide the number of moles of sodium bicarbonate by this total. This is the mole fraction of sodium bicarbonate. You will notice that the mole fraction increases from Experiment 1 through Experiment 7. Enter these data in Table 2.

$V = C^3 / 6\pi^2$

4. Calculate the volume of the balloon from the circumference. Recall that the circumference of a circle is $2\pi r$, and the volume of a sphere is $(4/3)\pi r^3$. So take the average circumference of the balloon (in cm) you measured, cube this circumference, and divide by $6\pi^2$ (= 59.2). The result is the volume of the balloon. Enter the volumes in Table 2.

5. Draw a graph of the mole fraction of sodium bicarbonate used (on the x axis) *vs.* the volume of carbon dioxide produced (on the y axis). Include the zero points at *both* ends of the graph (these are included in Table 2) so you will plot a total of 9 points. See Figure 2. Draw a best-fit straight line through the first few points and another best-fit straight line through the last few points. The point where the lines cross represents the stoichiometry of the reaction.

6. Determine the mole fraction of sodium bicarbonate at the point at which the lines cross and enter the value in Table 3 on your data sheet. Comment on this value; is it what you expected?

7. Enter your data into the lab computer. The program will check your calculations and plot a graph based upon what it calculates to be the best straight lines through your data points. (The program uses a least squares analysis of the data.) Record the result in Table 4 on your data sheet.

Calculation of the volume of carbon dioxide

1. The volume of carbon dioxide produced is determined by measuring the circumference of the balloon. The number of moles of carbon dioxide produced can be calculated from this volume using the ideal gas law.

The barometer in your laboratory is calibrated in hPa (hectpascals). The relationship between hPa and atm is:

$hPa \times 9.869 \times 10^{-4} = atm$

2. Obtain current readings of the atmospheric pressure and room temperature from your instructor and record these values in Table 5.

3. Choose one of your experimental points near the intersection of the two lines where the volume of carbon dioxide is the greatest. Read *from your graph* the volume of carbon dioxide corresponding to this mole fraction of sodium bicarbonate. That is, do *not* record the actual volume measured, but the volume according to the best straight line you have drawn through the points. See Figure 2. Also for this point, decide which reactant is the limiting reactant and note this, and its number of moles, in Table 5.

Figure 2. Plot of carbon dioxide produced
vs. mole fraction of sodium bicarbonate

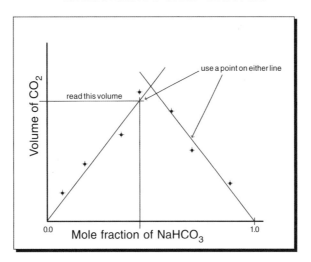

4. The number of moles of carbon dioxide can now be calculated using the ideal gas law:

Pressure × Volume = Number of moles × R (gas constant) × Temperature

The gas constant R = 0.08206 L atm/mole K

$$\text{The number of moles of } CO_2 = \frac{\text{Pressure} \times \text{Volume}}{R \times \text{Temperature}}$$

5. Compare the two molar quantities; is the result what you expected? Comment in Table 5.

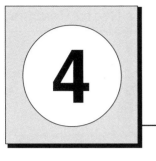

The Stoichiometry of a Redox reaction

Objectives

In this experiment you will determine the stoichiometry of another chemical reaction—in this case a redox reaction. You will investigate the oxidation of iron by copper sulfate. The stoichiometry of the reaction will be determined from the mass of iron oxidized and the mass of copper reduced. The amounts of metal are very small and it is essential that you do the experiment very carefully with high precision and accuracy.

Oxidation-Reduction Reactions

There are two principal types of inorganic reactions: acid-base reactions such as the one between vinegar and baking soda investigated in Experiment 3 and oxidation-reduction (redox) reactions such as the one investigated in this experiment.

Wood burning in the fireplace, iron rusting, and zinc reacting with hydrochloric acid to form hydrogen gas, are all examples of *oxidation-reduction* or *redox* reactions. In these reactions a movement of electrons from one reactant to another is taking place—one element loses electrons and another element gains electrons. The loss of electrons is called *oxidation* and the gain of electrons is called *reduction*.

A useful mnemonic is:

OIL   Oxidation is loss
RIG   Reduction is gain

At one time oxidation referred to the combination of an element with oxygen to form an oxide and reduction meant the removal of oxygen from an oxide to form the element. For example, the iron age commenced when it was learned that charcoal could be used to reduce iron ore (iron(III) oxide Fe_2O_3) to iron metal. Now the definitions of oxidation and reduction are much broader and describe the removal or addition of electrons in any reaction.

In every redox reaction, some substance is oxidized and another substance is reduced—one cannot occur without the other. When a substance is oxidized it loses electrons. Where do the electrons go? They must be acquired by some other substance that is reduced in the reaction.

The substance that is oxidized reduces the other substance—and therefore it is called the *reducing agent*. Likewise, the substance that is reduced is by definition the substance that oxidizes the other—it is called the *oxidizing agent*. Ask yourself, if a substance is reduced, from what substance is it getting the electrons? They must come from the substance that is being oxidized. Quantitatively, the number of electrons lost by one substance in the reaction must equal the number of electrons gained by another substance in the same reaction.

Transition metals, such as iron, are characterized by a multiplicity of oxidation states. For example, in its compounds, iron can exist in the +3 oxidation state as the ion Fe^{3+} or in the +2 oxidation state as the ion Fe^{2+}. Copper exists most commonly in a +2 oxidation state as Cu^{2+}, but in several compounds it has an oxidation state of +1. The object of this experiment is to determine the oxidation state of the iron produced in the reaction—is it Fe^{2+} or Fe^{3+}?

Procedure

Work with a partner for this experiment but prepare and submit individual data sheets. Write the name of your partner on your data sheet.

The reason for using pairs of nails is to increase the surface area and the rate of the reaction.

You might want to practise with a spare nail in another test tube while the reaction is proceeding.

Do not get the copper(II) sulfate solution on your skin. If you do, wash it immediately with water.

Dispose of the copper sulfate solution properly—follow the directions of your lab instructor.

Dispose of the methanol and acetone properly—follow the directions of your lab instructor.

Fe: 55.85

Cu: 63.55

1. Weigh two clean dry 6" test tubes and make a note of their masses in Table 1. Your partner should do the same and will duplicate the entire experiment.

2. Add 1M copper(II) sulfate solution to each test tube to a depth one-half inch less than the length of one of the iron nails.

3. Clean four iron nails with steel wool to remove any rust and contaminants. Wash the nails with water and dry thoroughly with paper towel. Divide the nails into two pairs. Determine the masses of both pairs and record the results in Table 1 (to 3 decimal places). Use tongs to manipulate the nails and a clean beaker to carry them around in until the final weighing. This will avoid contamination.

4. Place a pair of nails in each test tube containing copper(II) sulfate solution so that the heads remain clear of the surface. Leave the nails in the solution for the next 25–30 minutes. Record in Table 2 any observations that would lead you to understand that some reaction is occurring.

5. Fashion a hook out of copper wire and remove one of the nails. Use a spatula to dislodge any copper metal sticking to the nail into the test tube. Wash off any copper on the nail using a distilled-water wash bottle. It's essential not to lose any copper! Repeat this procedure with the second nail. Then repeat again for the second test tube.

6. Dry the nails using paper towel and reweigh. Record the masses in Table 1 on your data sheet. Calculate the mass lost by each pair of nails through oxidation.

7. Carefully decant the supernatant liquid from a test tube into a 100 mL beaker. Do this carefully so that none of the copper particles are lost. If, by accident, some copper is decanted with the liquid, it must be returned to the test tube and the decanting procedure should be repeated. Add a small amount (<5 mL) of methanol to the solid copper product and swirl gently to wash the material. Decant the methanol into another beaker, again making sure not to lose any product. Repeat using another <5 mL of methanol and then two <5 mL portions of acetone. Gently warm the copper metal in the test tube in a hot water bath on a hot plate to vaporize any remaining acetone. Remove the test tube and allow it to cool. The copper product should be dry and be able to move freely in the test tube. Repeat the entire procedure using the second test tube.

8. Make sure that the outside of the test tube is clean and dry and weigh it. Record the mass of the test tube and copper metal in Table 1. Reheat, cool, and reweigh the test tube. The mass should be the same; if not, repeat. Determine the mass of copper formed in the reaction. Repeat for the second test tube.

Calculations

1. Calculate the number of moles of iron oxidized by dividing the mass lost by the atomic mass of iron. Calculate the number of moles of copper produced by dividing the mass of copper by the atomic mass of copper. Then determine the mole ratio iron/copper. This ratio is the stoichiometric ratio of the reaction.

2. In Table 3 write the balanced equation for the reaction based upon your data.

Chemical Reactions and their Equations

Objectives

In this experiment you will investigate various chemical reactions in aqueous solution and learn how to characterize them. You will also learn how to write the chemical equations representing the reactions. From the overall equations you will derive the detailed ionic and net ionic equations representing the reactions.

Introduction

In a chemical reaction, atoms are rearranged—bonds between atoms are broken and new bonds between atoms are formed. Since these processes are not directly observable, how do we know when a chemical change has occurred? In a reaction, one or more substances, the reactants, are changed into one or more new substances, the products and the formation of a new substance is evidence that a reaction has occurred. This might be the formation of a precipitate, the evolution of a gas, or a change in the color of the system.

Since atoms are neither created nor destroyed in a chemical reaction, the number and identity of the atoms at the end of a reaction must be the same as the number and identity of the atoms at the beginning. This is why the symbolic representation of a chemical reaction is called a chemical equation—both sides are equal.

The equation representing a reaction is developed by first writing down the formulas for all the substances involved. The reactants are written on the left side and the products are written on the right side. An arrow written between the two sides indicates the progress of the reaction from reactants to products:

$$\text{Reactants} \quad \rightarrow \quad \text{Products}$$

Note that the substances are written in the form in which they exist. If necessary the states of the various participants in the reaction can be included. This is done by following each substance with a notation *(g)*, *(l)*, *(s)*, or *(aq)* indicating gas, liquid, solid, and aqueous solution respectively. For example, the equation representing the oxidation of aluminum is:

$$4Al\textit{(s)} \quad + \quad 3O_2\textit{(g)} \quad \rightarrow \quad 2Al_2O_3\textit{(s)}$$

Numbers called stoichiometric coefficients are placed before the formulas for the substances to balance the equation. This ensures that the equation obeys the law of the conservation of mass—the numbers of atoms on the two sides of the equation are equal. The formulas themselves must not be changed because that would change the identities of the substances they represent. The ratio of the coefficients is called the mole ratio, or stoichiometry, of the reaction.

Solutions, solutes, and solubility

A solution is a homogeneous mixture of two or more substances. To make a solution, one substance is dissolved in another substance. However, not all substances are able to dissolve. If the substance does dissolve it is *soluble*; if it doesn't, it is *insoluble*. Some substances dissolve but only to a limited extent. The *solubility* is the maximum amount of substance that will dissolve in a certain quantity of water at a certain temperature.

The substance that is dissolved is called the *solute*. The substance the solute is dissolved in is called the *solvent*. It is the solvent that determines the physical state of the solution. We will be concerned specifically with solutions in which the solvent is water. Such solutions are called aqueous solutions.

If a solute produces ions when it dissolves, it facilitates the passage of an electric current though the solution and is called an *electrolyte*. Some solutes produce no ions when they dissolve and are referred to as *nonelectrolytes*. Electrolytes themselves are divided into two categories. Those electrolytes that produce a high concentration of ions because the solute breaks up completely into ions are called *strong* electrolytes. Other solutes produce some ions when they dissolve but remain largely unionized in solution and these are called *weak* electrolytes.

Types of reactions in aqueous solution

Reactions in aqueous solution can be divided into two principal types: acid-base reactions and reduction-oxidation reactions. It is useful to characterize two additional classes of reactions. These are reactions in which a gas or precipitate is formed and we'll look at these in a moment. First let's examine acid-base reactions.

When an acid reacts with a base in aqueous solution, a salt and water are formed. The reaction is called *neutralization*. An example is the reaction of vinegar with sodium bicarbonate you investigated in an earlier experiment. Another example is the reaction between solutions of hydrochloric acid and sodium hydroxide to form sodium chloride and water:

$$HCl(aq) \ + \ NaOH(aq) \ \rightarrow \ NaCl(aq) \ + \ H_2O(l)$$

There are seven common strong acids: hydrochloric acid HCl, hydrobromic acid HBr, hydroiodic acid HI, nitric acid HNO_3, sulfuric acid H_2SO_4, chloric acid $HClO_3$, and perchloric acid $HClO_4$. The last four of these are called oxoacids.

Hydrochloric acid, sodium hydroxide, and sodium chloride are all strong electrolytes and it is instructive to write these substances in the form in which they exist in solution, that is, as ions:

$$H_3O^+(aq) \ + \ Cl^-(aq) \ + \ Na^+(aq) \ + \ OH^-(aq) \ \rightarrow \ Na^+(aq) \ + \ Cl^-(aq) \ + \ 2H_2O(l)$$

This equation is called a *detailed ionic equation*. All the ions present are listed. Notice, however, that the sodium ion $Na^+(aq)$ occurs on both sides of the equation. In other words, nothing happens to the sodium ion when the solutions of hydrochloric acid and sodium hydroxide are mixed. It is not involved at all in the chemical reaction and for that reason is called a spectator ion. The chloride ion $Cl^-(aq)$ is another spectator ion. When the spectator ions are deleted from both sides of the equation, the result is the *net ionic equation*:

$$H_3O^+(aq) \ + \ OH^-(aq) \ \rightarrow \ 2H_2O(l)$$

This equation represents what happens in solution when any strong acid reacts with any strong base. The only difference when the acid or base is changed is the salt that is formed in the reaction. The net ionic equation illustrates that the driving force behind the neutralization of any acid by any base in aqueous solution is the formation of the weak electrolyte water.

The key to developing a successful net ionic equation is to write all the substances involved in the form in which they predominantly exist in aqueous solution. This means writing strong electrolytes as ions and weak electrolytes as molecules or formula units. Gases are written as molecules and insoluble substances are written as molecules or formula units.

Precipitation reactions

General solubility rules are listed in Table 1.

Some salts are only very sparingly soluble in water. Silver chloride, AgCl, is an example of an sparingly soluble salt. What happens then when a solution of a salt containing the silver ion is mixed with a solution of another salt containing the chloride ion? Because the silver chloride is essentially insoluble, the silver chloride precipitates from the mixed solutions. Consider, for example, mixing a solution of silver nitrate with a solution of sodium chloride:

$$AgNO_3(aq) + NaCl(aq) \rightarrow AgCl(s) + NaNO_3(aq)$$

The (s) indicates the formation of the precipitate—the insoluble salt. A clearer picture is provided when the substances involved are written in the form in which they predominantly exist in aqueous solution. All the salts in solution are written as separate ions—they are strong electrolytes. The solid precipitate is written as the solid insoluble salt—it is not in solution:

$$Ag^+(aq) + NO_3^-(aq) + Na^+(aq) + Cl^-(aq) \rightarrow AgCl(s) + Na^+(aq) + NO_3^-(aq)$$

This is a detailed ionic equation. There are spectator ions in the equation that can be cancelled to derive the net ionic equation:

$$Ag^+(aq) + Cl^-(aq) \rightarrow AgCl(s)$$

Table 1. Solubility Rules

Soluble salts:

all salts of Na^+, K^+, and the ammonium ion NH_4^+

all NO_3^-, $CH_3CO_2^-$, ClO_3^-, and ClO_4^-

Cl^-, Br^-, and I^-	except Ag^+, Hg_2^{2+}, Pb^{2+}, and Cu^+
F^-	except Mg^{2+}, Ca^{2+}, Sr^{2+}, Ba^{2+}, and Pb^{2+}
SO_4^{2-}	except Pb^{2+}, Sr^{2+}, and Ba^{2+}
	(Ag_2SO_4, Hg_2SO_4, $CaSO_4$ are only slightly soluble)

Insoluble salts:

CO_3^{2-}, CrO_4^{2-}, $C_2O_4^{2-}$, and PO_4^{3-}	except Na^+, K^+, and NH_4^+
OH^- and O^{2-}	except Na^+, K^+, Rb^+ etc, Ba^{2+} and NH_4^+
	($Ca(OH)_2$ and $Sr(OH)_2$ are slightly soluble and $Mg(OH)_2$ is only very slightly soluble)
S^{2-}	except Group 1A, Group 2A, and NH_4^+

The net ionic equation indicates the formation of the insoluble silver chloride from silver ions and chloride ions in solution. It does not matter where the silver ions come from; it could be silver acetate, silver nitrate, silver perchlorate, or any other soluble silver salt. Nor does it matter where the chloride ions come from, sodium chloride, potassium chloride, or any other soluble chloride salt. The result is the same—the precipitation of the insoluble silver chloride. The precipitation of the insoluble salt, and therefore its removal from the solution, drives the reaction to completion.

Gas-forming reactions

The analysis of reactions in which a gas is formed is similar. Consider for example the formation of carbon dioxide in the reaction between magnesium carbonate and nitric acid:

$$2HNO_3(aq) + MgCO_3(s) \rightarrow Mg(NO_3)_2(aq) + H_2O(l) + CO_2(g)$$

The detailed ionic equation is:

$$2H_3O^+(aq) + 2NO_3^-(aq) + MgCO_3(s) \rightarrow Mg^{2+}(aq) + 2NO_3^-(aq) + 3H_2O(l) + CO_2(g)$$

The nitrate ions are spectator ions and can be deleted from both sides of the equation. So the net ionic equation is:

$$2H_3O^+(aq) + MgCO_3(s) \rightarrow Mg^{2+}(aq) + 3H_2O(l) + CO_2(g)$$

The net reaction is the same for any strong acid. The production of the gas, and therefore its removal from the solution, drives the reaction to completion.

Oxidation-reduction reactions

As described in an earlier experiment, in a oxidation-reduction (redox) reaction, electrons are transferred from one reactant to another. One element loses electrons and is oxidized and another element gains electrons and is reduced. In every redox reaction, some substance is oxidized and another substance is reduced—one cannot occur without the other. Quantitatively, the number of electrons lost by one substance must equal the number of electrons gained by another substance.

A practical method you can use to differentiate between acid-base reactions and redox reactions is to look for a change in the oxidation number of any element involved. If no oxidation numbers change then the reaction is an acid-base reaction. If there is some change, then the reaction is a redox reaction. Oxidation numbers can be worked out from first principles if the structure and bonding in the compound is known. However, until you know more about the electronic structures of molecules and polyatomic ions, it is best to follow the few straightforward rules for the determination of oxidation numbers listed in Table 2.

Procedure

Work with a partner but prepare and submit individual data sheets. Write the name of your partner on your data sheet.

Precipitation reactions

1. Get from your instructor a sheet of Saran® wrap, a white sheet of paper, a black sheet of paper, and a box of eight solutions. These solutions are provided in small plastic pipets and contain the four cations and the four anions to be tested. When you use the pipets, hold them in a nearly vertical position and dispense *only two* drops. When you add drops of a second solution to the first set of drops *do not allow* the pipette tip to touch the first drop.

Table 2. Rules for the Determination of Oxidation Numbers

1. An atom of any element by itself has an oxidation number of zero.

2. For a monatomic ion, the oxidation number is the charge on the ion.

3. For a polyatomic ion, the sum of the oxidation numbers of all the participating atoms equals the charge on the ion.

4. In compounds, fluorine always has an oxidation state of –1.

5. In compounds, oxygen always has an oxidation state of –2, except when combined with fluorine alone, or in peroxides and superoxides.

6. Halogens Cl, Br, and I, have oxidation numbers of –1, except when combined with oxygen or fluorine (or another halogen higher in the group).

7. Hydrogen has an oxidation number of +1 when combined with non-metals, and an oxidation number of –1 when combined with metals.

8. In compounds, Group 1A elements, the alkali metals, always have an oxidation number of +1. In compounds, Group 2A elements, the alka line earth metals, always have an oxidation number of +2.

$AgNO_3$ *(aq)*, source of Ag^+ *(aq)* Na_2SO_3 *(aq)*, source of SO_3^{2-} *(aq)*

$Cu(NO_3)_2$ *(aq)*, source of Cu^{2+} *(aq)* KI *(aq)*, source of I^- *(aq)*

$Co(NO_3)_2$ *(aq)*, source of Co^{2+} *(aq)* Na_3PO_4 *(aq)*, source of PO_4^{3-} *(aq)*

$Ni(NO_3)_2$ *(aq)*, source of Ni^{2+} *(aq)* $NaHCO_3$ *(aq)*, source of HCO_3^- *(aq)*

2. Draw the grid shown in Table 3 (data sheet) on the white paper provided. Place the white paper on top of the black paper and place the Saran® wrap over the grid. Dispense the drops in the appropriate squares. Before recording your observations wait 2-3 minutes to ensure that the reaction is complete. Slide the white paper out and check the colors against a black background.

3. Record your observations in Table 3 on your data sheet. Remember that you are looking for evidence of precipitate formation—merely adding a colored solution to a colorless solution does not indicate a chemical reaction. If no reaction occurs, write "NR" in that square.

4. Return all pipettes to the appropriate container with the *tip pointing up* when closed. All plastic wrap should be rinsed off, folded up, and placed in the designated waste container.

5. Your instructor will assign two reactions at random—two combinations of two solutions in which there were precipitates. In Table 4 on your data sheet write

the overall chemical equation for each reaction, then write the detailed ionic equations, and finally derive and write the net ionic equations for each.

Acid-base reactions

1. Prepare the following solutions:

   Add a few drops of 0.1M HCl to a small test-tube and then add 1-2 drops of phenolphthalein, an example of an acid-base indicator. Note the color and record in Table 5 on your data sheet.

   Add a few drops of 0.1M NaOH to a second test tube, followed by 1-2 drops of phenolphthalein. Note the color and record in Table 5.

Make sure that you use the correct HCl solution: 0.1M not 6.0M.

2. Add ten drops of 0.1M HCl to a small test tube followed by 1-2 drops of phenolphthalein indicator. Then, using a pipet of the same type, begin adding drops of 0.1M NaOH, being careful to keep track of the number of drops added. Shake the test tube after each drop of NaOH is added and keep adding drops until the solution shows the slightest persistent color change. Repeat this twice more and record the average number of drops of 0.1M NaOH needed when you used 10 drops of 0.1M HCl. Record these results in Table 5.

3. Repeat the above experiment, substituting 0.1M H_2SO_4 for the 0.1M HCl. Record the average of the three trials in the table.

4. For each of the above experiments in steps 2 and 3 write:

   a complete molecular equation.
   a detailed ionic equation.
   a net ionic equation.

5. Reconcile the results of the two experiments and the ionic equations you wrote for each experiment. Explain the results.

Gas-forming reactions

1. The following laboratory tests are used to identify the following gases:

   Hydrogen: Test with a burning splint in an inverted test tube.
   Result: Popping noise as the hydrogen ignites.

   Oxygen: Test with a glowing splint with the mouth of the test tube tilted slightly from the vertical and pointed upwards.
   Result: Splint bursts into flame.

   Carbon dioxide: Test with a glowing splint and hold the test tube vertical with the mouth pointed upward.
   Result: Flame is extinguished.

Exercise caution when testing these gases by making sure that the mouth of the test tube is not pointing directly at yourself or at anyone else when you test the gas.

2. Record the results of each of the following experiments in Table 6 on your data sheet and write **balanced** chemical equations describing what happened.

3. Add a small chip of marble to a 13×100mm test tube. Then add 15-20 drops of 6M HCl and wait for at least 30 seconds. Insert a lit wooden splint about half-way into the test tube and observe the results.

The KI in this reaction is a catalyst. One of the products of the reaction is water.

4. Add 15-20 drops of 3% hydrogen peroxide to a 13×100mm test tube followed by a crystal of solid potassium iodide KI. Wait 45 seconds. Insert a glowing wooden splint about halfway into the test tube and observe the results.

5. Cut half of the stem and half of the base off a plastic pipet so that it will fit snugly on the top of a 13×100mm test tube. Place 15 – 20 drops of 6M HCl in the test tube. Place a piece of mossy zinc about the size of a pea in the bulb of the plastic pipet. Start the reaction by inverting the pipet and placing it over the test tube so that the zinc falls into the 6M HCl. Place a second test tube upside down over the stem of the pipet. After about 45 seconds, remove the top test tube (keeping it inverted) and carefully insert a lit splint into the end and observe the results.

Oxidation-reduction reactions

1. Add a small piece of zinc to a 13×100mm test tube filled about 1/3 full with 1M $CuSO_4$. Set this aside until you have finished all of your other experiments and then record your observations in Table 7 on your data sheet.

2. Add a small piece of copper to a 13×100mm test tube filled about 1/3 full with 1M $ZnSO_4$. Set this aside until you have finished all of your other experiments and then record your observations.

This is a confirmation test for molecular bromine.

3. Add 20 drops of bromine water to a 13×100mm test tube. Then add 15 drops of baby oil. Cork the test tube and shake by gently tapping the test tube on the side with your finger. Note the color imparted to the mineral oil by the bromine molecules in Table 7.

The purpose of adding the acid to the bleach is to increase the production of chlorine Cl_2 gas.

4. Add 30 drops of 0.2 M potassium bromide to another test tube. To this add 15 drops of baby oil, then 10 drops of bleach, followed by 2-3 drops of 6 M HCl. Shake as described in Step 3. Record your observations in Table 7.

This is a confirmation test for molecular iodine.

5. Place 1-2 small crystals of solid iodine in another test tube. Add 15 drops of baby oil and shake by gently tapping with your finger as described in Step 3. Note the color imparted to the mineral oil by the iodine I_2 molecules.

6. Add 30 drops of 0.1 M potassium iodide to another test tube. To this add 15 drops of baby oil, then 10 drops of bleach, followed by 2-3 drops of 6 M HCl, and shake. (Remember that Cl_2 gas is released when acid is added to bleach).

7. To another test tube, add 20 drops of 0.1 M potassium iodide, 12–15 drops of baby oil, and finally 6–7 drops of bromine water. Shake by gently tapping with your finger as described in Step 3. Record your observations in Table 7.

8. For each step 1-7 write the balanced molecular equation representing what you observed. If no reaction took place, write "NR" on the product side of the equation.

11:30

The Heat Capacity of Metals

Objectives

In this experiment you will determine the heat capacities of various metals by calorimetry and learn something about heat, temperature, and energy. You will determine the average atomic weight of an unknown alloy using the Law of Dulong and Petit.

Introduction

When a system is heated, energy is transferred into the system from its surroundings. This results in an increase in the motion of the atoms and molecules within the system. This increase in motion can take several forms. If, for example, the system consists of gas molecules, then the molecules can move about more quickly; this is an increase in *translational energy.* Similarly, atoms within the individual molecules can vibrate about their positions more vigorously; this is an increase in *vibrational energy.* Finally, the molecules in a gaseous system may rotate more rapidly; this is an increase in *rotational energy.*

However, if the system is solid, the atoms or molecules are packed close together and they cannot move relative to one another. An increase in individual translational energy is impossible. Similarly, rotation of the atoms or molecules held within the crystal lattice is impossible. All that can happen when a solid—a metal, for example—is heated is that the atoms vibrate more within the crystal lattice.

The **heat capacity** of an object is the quantity of heat required to raise the temperature of the object one degree Celsius (or Kelvin). Units: J K^{-1}

It may not be surprising then to discover that the ability of a metal to absorb heat energy (its *heat capacity*) depends not upon the *mass* of its atoms but rather upon the *number* of atoms within the sample. In other words, the heat capacity *expressed as a molar quantity* should be the same for all metals.

This law was recognized by Pierre L. Dulong and Alexis T. Petit as part of a systematic study of the heat capacities of various metals. In 1819 they published their observation that the atomic mass of a metal times its heat capacity per gram seemed to be a constant equal to 5.96 cal K^{-1}mol^{-1} or 24.9 JK^{-1}mol^{-1}. The Law of Dulong and Petit was used in the nineteenth century by Berzelius and later by Cannizzaro to determine the atomic masses of numerous metals.

The **joule** is the SI unit for work or energy A **calorie** is the quantity of heat required to raise the temperature of one gram of water one degree Celsius (or Kelvin).

1 cal = 4.184 J (exactly).

It was not until 1907 that the law of Dulong and Petit was explained by Albert Einstein. He showed that the heat capacity of a solid is zero at 0 K and increases as the temperature increases to a maximum value of three times the gas constant R. See Figure 1. Provided that the temperature is sufficiently high, and room temperature is high enough, this means that all metals have the same molar heat capacity of 3R.

The heat involved in a chemical reaction or in a physical process can be determined using a device called a *calorimeter.* As its name suggests, a calorimeter is used to measure heat. The calorimeter that you will use in this experiment is illustrated in Figure 2. It consists of two nested polystyrene coffee cups.

Figure 1. Molar heat capacity as a function of temperature

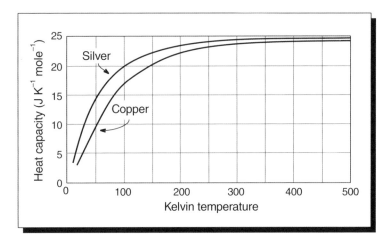

A polystyrene cup is an excellent insulator. As you may have experienced, you cannot feel any heat through the polystyrene even when the cup is full of very hot coffee. Because of this efficient insulation, we can assume that a reaction taking place inside a coffee-cup calorimeter is completely isolated from its surroundings outside. We can also assume that the heat absorbed by the calorimeter itself is zero; *i.e.,* that the heat capacity of the coffee-cup calorimeter is zero.

The name calorimeter comes from
calor: heat + *meter:* measure

Although the heat involved in the chemical reaction cannot be measured directly, it can be determined indirectly by measuring the temperature change. The heat change and the temperature change are related by a property of the system called *heat capacity.* The heat capacity is the heat required to raise the temperature of the system by one degree:

Figure 2. Calorimeter made from two nested plastic cups

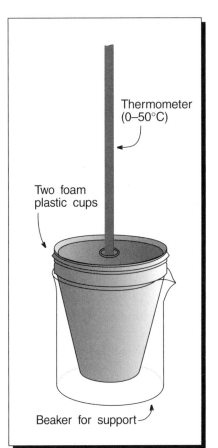

Thermometer (0–50°C)

Two foam plastic cups

Beaker for support

$$\text{Heat} = \text{heat capacity} \times \text{temp change}$$
$$\text{J} = \text{JK}^{-1} \times \text{K}$$

According to the first law of thermodynamics, energy is conserved. In a calorimeter isolated from the outside world, if one substance loses heat, then another substance must gain an equal amount of heat:

$$\text{Heat gained} = \text{heat lost}$$

Furthermore, according to the second law of thermodynamics, it is always the hotter substance that loses heat and the colder substance that gains heat. In other words, heat always travels in the direction from hot to cold.

In this experiment, a hot block of metal is immersed in cold water in the calorimeter. The metal loses heat and the water gains an equal amount of heat. The temperature of the metal decreases and the temperature of the water increases until both are at the same temperature. By monitoring the temperature change, it is possible to calculate the heat capacity of the metal.

Heat capacity is an extensive property; it depends upon the size of the sample. For example, it takes twice as much heat to raise the temperature of 2.0 liters of water by 1°C as it does to raise the temperature of 1.0 liter of water by 1°C.

The **specific heat** of a substance is the heat capacity of one gram of the substance. In other words, it is the quantity of heat required to raise the temperature of one gram of substance one degree Celsius (or Kelvin).

Units: $JK^{-1}g^{-1}$.

Specific heat, on the other hand, is an intensive property; it describes the heat capacity of *one gram* of a substance:

Heat capacity = specific heat × mass

$$JK^{-1} \quad = \quad JK^{-1}g^{-1} \times \quad g$$

The molar heat capacity is also an intensive property; it describes the heat capacity of one mole of the substance.

Heat capacity = molar heat capacity × moles

$$JK^{-1} \quad = \quad JK^{-1}mol^{-1} \quad \times \; mol$$

In this experiment you will determine the heat capacities of aluminum, copper, and lead blocks. You will also measure their masses and therefore will be able to calculate their specific heats. Then, using the law of Dulong and Petit, you will calculate the average molar mass of an unknown alloy.

Procedure

Work with a partner for this experiment but prepare and submit individual data sheets. Write the name of your partner on your data sheet.

Start with the hot plate full on and reset to 6 to simmer.

1. Fill a 1000 mL beaker three-fourths full with distilled water and heat it on a hot plate to a gentle simmer (hot plate setting 6). Add five or six boiling chips to prevent bumping.

2. While the water heats, weigh your four blocks of metal: aluminum, copper, lead, and an unknown alloy. Record the masses in Table 1 on your data sheet.

3. Place the four metal blocks into the beaker of simmering water. The metal blocks must remain in boiling water for at least ten minutes to ensure that they reach 100°C. Use the tongs to hold the metal blocks. Be very careful not to drop the the blocks into the beaker. It will break!

4. Assemble a coffee-cup calorimeter as shown in Figure 2. Place two foam plastic cups one inside the other and support the cups inside a 400 mL beaker. Fit a plastic lid and insert a 0–50°C thermometer supported by a clamp on a stand.

5. Measure *exactly* 100 mL of distilled water in a graduated cylinder and pour it into the calorimeter.

6. Monitor the temperatures of the water in the calorimeter at one minute intervals. Record the readings in Table 2. At a predetermined time (preferably the six-minute mark), use tongs to take one of the metal blocks from the beaker and place it quickly into the calorimeter. Replace the lid and very gently swirl the calorimeter a few times; then allow it to stand. Continue to record the calorimeter water temperature at one minute intervals. It takes two or three minutes for the

system to reach equilibrium and it is important to continue to take temperature readings until they begin to decrease linearly as shown in Figure 3. Plot the data on the graph as you take the readings.

Figure 3. Time-temperature calorimeter data for metal samples

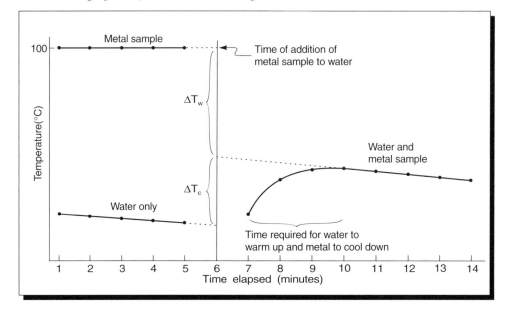

You must do several calculations and prepare five graphs. To use your time most efficiently, plot the graphs as the data are taken.

7. Repeat this routine using the other blocks of metal, including the unknown alloy. Enter all readings in Table 2 and plot the temperature readings on graphs like that in Figure 3. Assume that the temperature of the metal blocks at the time of mixing is 100°C. Extrapolate the lines to the point of mixing and determine ΔT_w and ΔT_c for each metal. Enter these values in Table 3.

8. Calculate the heat gained by the cold water in each experiment. Follow the calculations in Table 3. Then calculate the heat capacities of each metal block in units of JK^{-1}. Record the data in Table 3.

Using this graphical method, it is possible to extrapolate the experimental data to the point of mixing at which no measurement was actually taken. Use a ruler!

9. Calculate the specific heat of each metal and record the results in Table 3. Calculate the reciprocal of each specific heat. Plot a graph of the reciprocal of the specific heat against atomic mass for each of the three known metal blocks and draw the best straight line through the points. The origin of the graph should be used as one of the points (Figure 4).

Figure 4. Typical graph of molar mass (atomic weight) *vs.* 1/specific heat

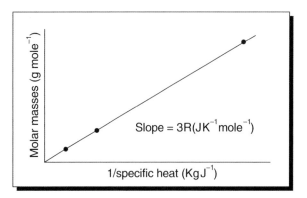

$$\text{Slope} = \frac{\text{g mole}^{-1}}{J^{-1}\,\text{Kg}}$$

$$= JK^{-1}\text{mol}^{-1}$$

The accepted value for the gas constant R is 8.3145 $JK^{-1}\,mol^{-1}$.

10. Plot the reciprocal of the specific heat of the unknown alloy and determine its average atomic mass from the graph.

11. From the slope of your graph for step 9, calculate the value of the gas constant R in $JK^{-1}mol^{-1}$. Determine the slope and then divide by three. Record the data in Table 4.

The Heat of Vaporization of Liquid Nitrogen

Objectives

In this experiment you will determine how much heat is required to vaporize liquid nitrogen. The experiment is an application of the First Law of Thermodynamics—the law of conservation of energy. You will learn how the energies involved in a chemical reaction or physical process can be determined experimentally using a calorimeter.

Introduction

Thermochemistry is the study of the heat changes associated with chemical reactions and physical processes. As a reaction proceeds, heat may be absorbed or liberated and the heat content of the system may change.

You may wonder why the term *enthalpy* is used instead of *energy*. Enthalpy is more useful and easier to use for processes occurring under constant atmospheric pressure. It is a measure of the quantity of heat leaving or entering a system under constant external pressure without having to worry about any work that is done on or by the system.

Just like energy, enthalpy is a *state function;* it depends only upon the current state of the system and not upon its history.

ΔH, the change in enthalpy, depends only upon the difference between the initial and final states of the system and not upon the route taken between the two—it is described as *path-independent*.

The symbol ΔH is commonly used to represent the change in the heat content of the system. The word used to represent heat content is *enthalpy* and ΔH is referred to as the *enthalpy change*. Enthalpy changes may be heats of reaction, heats of solution, heats of combustion, heats of vaporization, etc., depending upon the particular process. They can be positive, negative, or occasionally zero.

Processes or reactions in which positive heat changes occur (ΔH greater than zero) are called *endothermic* reactions—heat is absorbed. Reactions in which negative heat changes occur (ΔH less than zero) are called *exothermic* reactions—heat is liberated.

The heat liberated or absorbed in some process may be determined using the device called a *calorimeter.* The calorimeter is insulated from its surroundings so that no heat can escape from or enter the system. In such a situation any heat liberated by one component in the system must be gained by another component in the system. That is, any heat lost must equal any heat gained within the calorimeter.

The heat lost or gained in a process is actually determined by measuring the temperature change. The more heat that is lost or gained, the greater the temperature change. The relationship between the heat change and the temperature change is determined by the heat capacity of the system:

Heat (J) = Heat capacity (JK^{-1}) $\times$ temperature change (K)

If the system is homogeneous, then the heat capacity may be expressed in terms of the specific heat of the system and the mass of the system:

Heat capacity (JK^{-1}) = Specific heat (JK^{-1}g^{-1}) $\times$ mass (g)

Nitrogen, N_2, is normally a gas. However, if cooled sufficiently, nitrogen will liquefy. The temperature at which nitrogen gas liquefies, or liquid nitrogen boils, is -196°C under atmospheric pressure. The low boiling point indicates that the intermolecular bonding between the nitrogen molecules is relatively weak—not much energy is required to break the molecules apart. In this experiment you will determine just how much energy is required to vaporize liquid nitrogen.

There is a rule concerning the heats of vaporization of liquids called Trouton's Rule. This rule states that the heat of vaporization divided by the boiling point of a liquid (in K) is equal to a constant called Trouton's constant. This constant actually isn't very constant but it does usually lie within the range 70 to 90 J K^{-1} mol^{-1}. For nitrogen the value is near the low end, about 72 J K^{-1} mol^{-1} indicating very little order in the liquid state—in other words molecules in liquid nitrogen are very disordered and behave quite chaotically. You can use this rule to estimate what the heat of vaporization of liquid nitrogen should be.

Procedure

Work with a partner for this experiment but prepare and submit individual data sheets. Write the name of your partner on your data sheet.

Handle the electronic thermometers with care. They are expensive. Under no circumstance put the thermometer in the liquid nitrogen.

Keep the scale set for °C and turn them off when not in use.

You must be careful with the liquid nitrogen. It is extremely cold and will cause severe frostbite. Do not perform any unauthorized experiments with the liquid nitrogen.

Rather than carry the liquid nitrogen back to your lab bench (during which more will evaporate), you might find it better to add the liquid nitrogen to the warm water immediately after removing it from the balance.

1. Heat 100 mL of distilled water in a 250 mL beaker to approximately 65°C. While the water is heating, determine the mass of two nested polystyrene coffee cups.

2. Then add the warm water to the polystyrene cup and determine the total mass. Subtraction of the two masses will allow you to determine the mass of water in the cup.

3. Place the polystyrene cups in a beaker as shown in Figure 1 so that the apparatus doesn't fall over.

Figure 2. Calorimeter made from two nested plastic cups

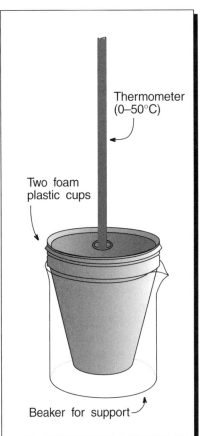

Thermometer (0–50°C)

Two foam plastic cups

Beaker for support

4. Determine the mass of another polystyrene coffee cup. Your instructor will then add about 80 mL of liquid nitrogen to this cup. Allow enough nitrogen to evaporate until about 60.0 g of nitrogen remains. Remember that you want 60.0 g of nitrogen so take into account the mass of the polystyrene cup. The mass doesn't have to be exactly 60.0 grams but you should determine the mass as precisely as possible.

5. While the nitrogen is evaporating down to 60.0 g, determine the temperature of the water in the first cup to the nearest 0.1°C.

6. *Remove the thermometer* and then quickly and carefully add the liquid nitrogen to the warm water. Observe and make a note of what happens. When the evaporation of the liquid nitrogen stops, or when you no longer hear a sizzling sound, fan away any remaining fog and swirl the water gently to melt any ice that may have formed. Do not reinsert the thermometer until these processes have finished.

7. Measure and record the temperature of the water.

8. Reweigh and record the mass of the two polystyrene cups and water. You will probably find that the mass of the water decreased during the experiment. In your calculations use the average of your two readings.

9. Repeat the experiment twice for a total of three runs. Be as precise as is possible in all your measurements. You should be able to determine, as a result of the three experiments, a value for the heat of vaporization of liquid nitrogen within 5 percent of the accepted value.

Calculations

The specific heat of water is 4.184 $JK^{-1}g^{-1}$.

1. Calculate the heat lost by the warm water (the enthalpy change ΔH) as it cooled to the final temperature:

ΔH = specific heat of water $\times$ mass of water $\times$ temperature change

2. Determine, therefore, the heat gained by the liquid nitrogen as it evaporated. First write on your data sheet the total heat gained; then calculate the heat gained per gram and the heat gained per mole N_2.

3. Calculate the heat of vaporization per mole N_2 using Trouton's Rule; how do the results of your experiment compare? Your instructor will tell you the established (literature) value for the heat of vaporization of liquid nitrogen.

Questions

1. Write an equation representing the vaporization of liquid nitrogen; include the value you have obtained for the heat of vaporization. Is the process exothermic or endothermic?

2. Identify two sources of error in this experiment that you believe might have led to an error in the value you have obtained. How could these errors be reduced?

3. When liquid nitrogen was added to the warm water a white fog was produced. What was this white fog?

4. Why was the mass of water less at the end of the experiment than at the beginning?

5. The heat of fusion (melting) of solid nitrogen is only 25.7 $J g^{-1}$. Why is this value so much lower than the heat of vaporization?

6. The heat of vaporization of water is 40,700 $Jmol^{-1}$. Why is this value so much higher than the heat of vaporization of liquid nitrogen?

The Shapes of Molecules and Ions

Objectives

In this experiment you will build models of simple molecules and polyatomic ions using valence shell electron pair repulsion theory. You will discover how the geometry of a molecule is determined by the most stable arrangement of the electron pairs in the valence shell of the central atom. Inspection of the molecular geometry will allow you to determine whether the molecule or polyatomic ion has a dipole moment.

Introduction

The way in which atoms combine together to form molecules has intrigued chemists for a long time. Why does one element have a particular affinity for another element, whereas two other elements do not react at all? Lavoisier, for example, could not explain why chlorine was so reactive whereas gold was not. From the middle of the 1800s, chemists began to think of the bonding between elements in terms of valence—a term developed by Frankland in 1852. The *valency* assigned to an element described its combining power—as illustrated in the stoichiometry of its compounds. The valency of an element was one of the properties that allowed Mendeleev to position elements logically in his Periodic Table. However, until Thomson identified the electron in 1897 the underlying reason behind valency could not be explained.

In the early 1900s Abegg, Drude, and later Thomson related the valency of an element to the number of outermost electrons in an atom. Thomson recognized that in the noble gases the outermost shell of electrons in the atom was complete—which suggested why they were unreactive. Meanwhile, Rutherford established the nuclear model for the atom in 1911 and Moseley established the magnitude of the nuclear charge and introduced the term *atomic number* in 1913. Then Niels Bohr in the same year introduced the principles of the quantization of energy in a description of how the electrons behave as they move around the nucleus.

It was in 1916 that Lewis published his ideas on chemical bonding although he had been using the approach since 1902. The basis of his theory was that chemical bonding takes place as a result of the sharing of electrons between two atoms—such that each atom, as a result of the sharing, attains an outermost group of eight electrons just like a noble gas. He represented this outermost group of electrons on the corners of a cube as shown in Figure 1. Bonding could occur by sharing electrons on an edge or a face. Shared electrons could originate from both atoms in a bond called *covalent* by Langmuir, or from just one of the atoms in a bond called *coordinate* by Sidgwick. One of the features of Lewis' theory that particularly appealed to chemists was the specific arrangement of electrons around the atom—an idea that could be used to explain why molecules had definite shapes.

The idea that electron pairs are arranged around an atom so as to minimize the repulsion between them was developed by Sidgwick and Powell in 1940. Their ideas were extended by Gillespie and Nyholm in 1957 and the approach became known

Figure 1. The Lewis-Langmuir concept of shared electrons

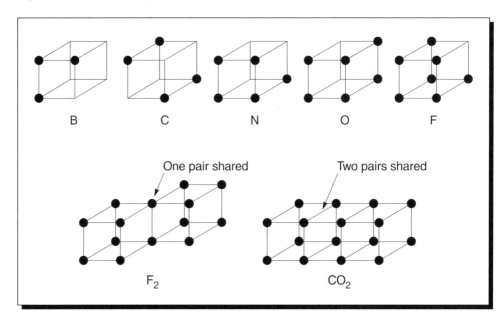

as *valence shell electron pair repulsion theory* or VSEPR. More recently, in 1992, Gillespie has modified the theory to explain the shapes of molecules through the mutual repulsion of different electron domains and the approach is sometimes referred to as *electron domain theory* or ED theory. VSEPR or ED theory is an empirical approach that allows the shape of a simple molecule or polyatomic ion to be determined.

Most chemists use the term Valence Shell Electron Pair Repulsion Theory, or VSEPR.

The shapes of molecules cannot be predicted on the basis of their formulas; for example, BeF_2 is linear in shape whereas water H_2O, with the same stoichiometry, is bent; BF_3 trigonal planar in shape but ammonia NH_3, with the same stoichiometry, is trigonal pyramidal?

In ED theory there are three different types of electron domains possible:

- A nonbonding or lone pair domain

- A single bond domain

- A double (or triple) bond domain

The basic principle is that electron domains are arranged so that they get as far away from each other as possible. The five basic minimum repulsion arrangements are shown in Figure 2.

The purpose isn't really to determine the arrangement of electron domains around the central atom in a molecule, it is to use this arrangement to derive the shape of the molecule. The shape of the molecule depends upon how many of the electron domains are nonbonding domains as shown in Figure 3. As an example, consider the molecule $SnCl_2$. In this molecule there are three electron domains around the Sn atom. So their arrangement is trigonal planar. However, only two domains are bonding (to the two Cl atoms) and the third domain is nonbonding. So the shape of the molecule—how the atoms are arranged—is V-shaped or bent.

Figure 2. The five basic VSEPR arrangements

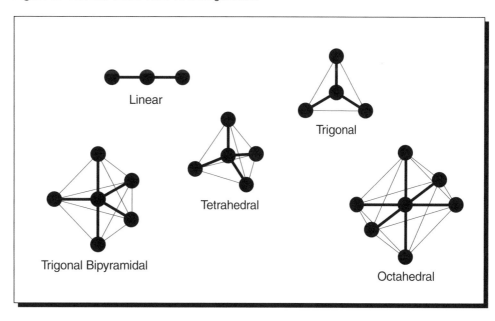

A nonbonding domain takes up the most amount of space around an atom because the electrons belong only to that atom and are not shared with another atom. A double (or triple) bond domain takes up more space than a single bond domain because it involves more pairs of electrons. So the arrangements shown in Figure 2 are often slightly distorted from the ideal geometries.

It isn't necessary to draw the Lewis electron dot structure in order to figure out the number of electron pairs around the central atom in a molecule. In fact, it is often more difficult to draw a good Lewis structure than determine the structure using VSEPR theory. In all simple molecules and polyatomic ions, the number of electrons in the valence shells of terminal atoms (except hydrogen) is always 8. So an easy method for the determination of the number of electron pairs around the central atom is to add up the total number of valence electrons in the molecule and divide by 8. The answer (the quotient) must equal the number of terminal atoms; the remainder is the number of electrons that reside on the central atom as lone (nonbonding) pairs.

If any of the terminal atoms are hydrogen, then dividing by 8 makes no sense since H has room for only two electrons in its valence shell. In this case the best approach is to subtract 2 from the total for each H before continuing with the calculation.

For example, consider again the molecule $SnCl_2$. The total number of valence electrons is 4 for the Sn plus 7 each for the two Cl, which equals 18. Dividing by 8 produces the quotient 2 which is the number of terminal Cl atoms. The remainder is 2, or one pair. So there are three pairs, or three electron domains, around the Sn atom—two bonding and one nonbonding. The arrangement is trigonal planar, and the shape of the molecule is bent or V-shaped.

The purpose of dividing by 8 is not to obtain the quotient because the quotient equals the number of terminal atoms. This is obvious from the formula of the compound.

The purpose is to obtain the remainder, in other words, to obtain the number of lone pairs of electrons on the central atom.

Once the molecular geometry is established, the polarity of the molecule can be determined. If the molecule is symmetrical, with each side or corner balanced by equal and opposite sides or corners, then the molecule will be nonpolar. However, if the molecule is asymmetric, with one side different from the opposite side, then the distribution of electrons will be unbalanced and the molecule will have a dipole moment.

Very often you will see the bonding between the atoms in a simple molecule described in terms of the *hybridization* of the orbitals on the central atom. The hybridization of the atomic orbitals is a method used in *valence bond theory* to produce a set of orbitals appropriate for the shape of the molecule. For example, in the $SnCl_2$ molecule described earlier, the electron domains (bonding and nonbonding) are arranged in a triangle around the Sn atom. However, the *s* and *p* orbitals do not point in the correct directions to accommodate these electron pairs—the molecule requires three orbitals on the Sn atom pointing to the corners of a equilateral triangle. The solution is to mix (or hybridize) the *s* and two of the *p* orbitals to produce a set of three hybrid orbitals called *sp²* hybrid orbitals. These hybrid orbitals point to the corners of an equilateral triangle and can be used to form the valence bonds with the two chlorine atoms and to accommodate the nonbonding electron pair. Three features of the hybridization process are important:

- The number of hybrid orbitals produced always equals the number of atomic orbitals used in the mix.

- The hybrid orbitals must accommodate all electron pairs, bonding and nonbonding. In other words the number of hybrid orbitals equals the number of electron domains.

- An appropriate set of atomic orbitals can be chosen to fit any observed arrangement of electron pairs around the central atom.

Common hybrid orbital sets and their geometries are:

sp	linear
sp²	trigonal planar
sp³	tetrahedral
sp³d	trigonal bipyramidal
sp³d²	octahedral

Recent calculations of molecular orbitals indicate that the use of d orbitals in simple molecules and polyatomic ions is minimal. So hybrid orbital sets involving d orbitals are not very realistic.

Procedure

You may work with a partner for the completion of Table 1 but you must complete Table 2 for your own set of cards on your own.

1. You will be provided with a molecular model set. In the set you will have sufficient atoms and bonds to make models of each of the five different arrangements. You will also be given a random set of eight formulas of molecules or polyatomic ions for which you will determine the shapes.

2. In Table 1 on the first page of your data sheet you will find a list of molecules or polyatomic ions. These molecules vary in the total number of electron domains around the central atom. For each molecule:

   - Determine which atom is at the center of the molecule—usually the least electronegative.

   - Determine the total number of valence electrons.

   - Divide this number by 8 to obtain the number of bonds around the central atom.

   - From the remainder, determine the number of nonbonding domains on the central atom.

   - Write the total number of electron domains.

   - Write the arrangement of electron domains and build the molecule.

   - State the required hybridization of atomic orbitals on the central atom.

Figure 3. Molecular shapes based upon the five basic arrangements

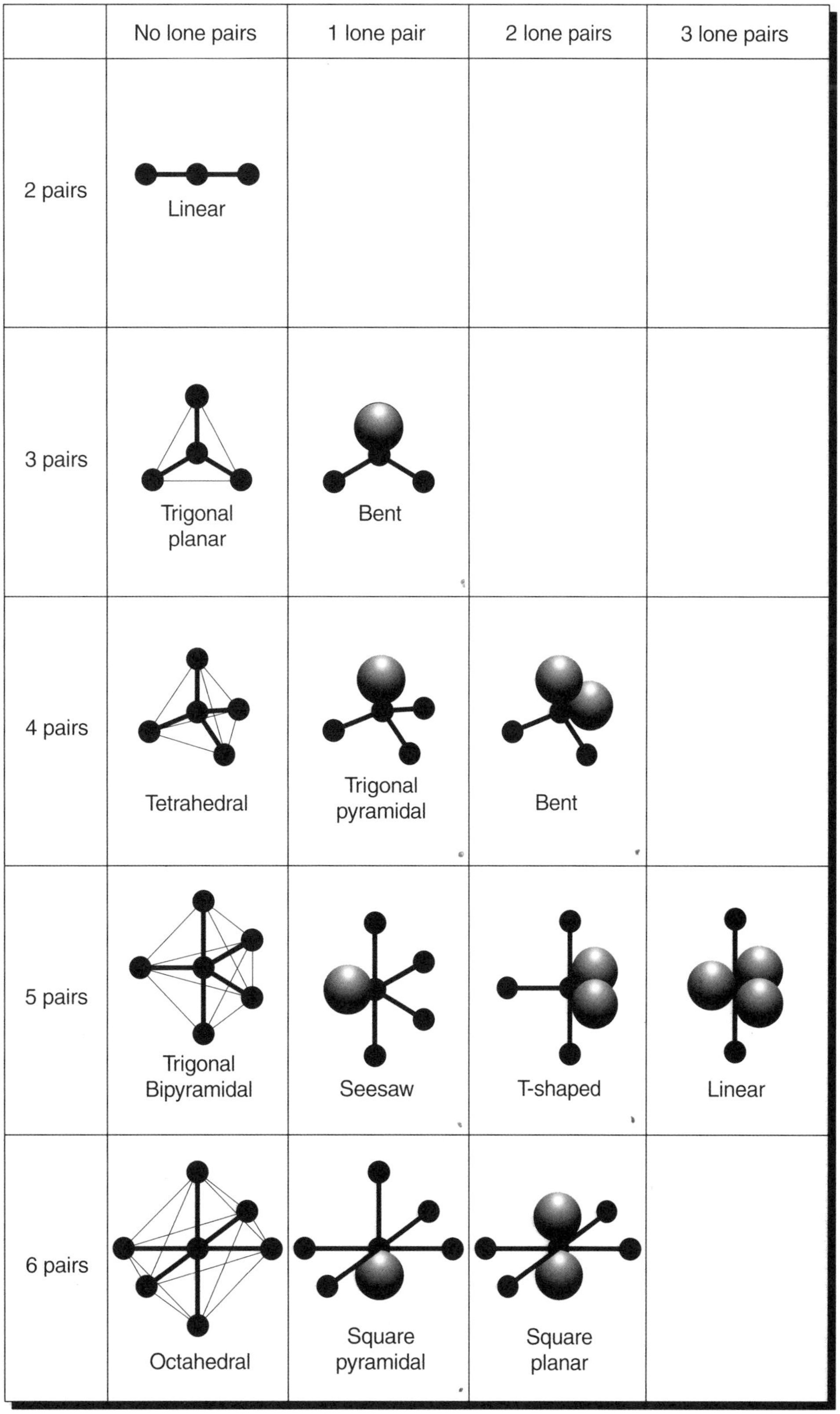

	No lone pairs	1 lone pair	2 lone pairs	3 lone pairs
2 pairs	Linear			
3 pairs	Trigonal planar	Bent		
4 pairs	Tetrahedral	Trigonal pyramidal	Bent	
5 pairs	Trigonal Bipyramidal	Seesaw	T-shaped	Linear
6 pairs	Octahedral	Square pyramidal	Square planar	

- Derive, describe, and draw the shape of the molecule.

- Determine whether or not the molecule or ion is polar.

Data for the first molecule is already completed for you. Check that the data shown is correct and follow the same organization for the other molecules in the table. You may collaborate with a partner in the lab in completing this table.

3. Repeat the procedure for the eight molecules or polyatomic ions listed on your own individual set of cards. You must show each completed model to your instructor who will initial your data sheet.

4. Answer the questions in Table 3.

5. Disassemble your models and check the set back in with your instructor.

The Molar Mass of Carbon Dioxide Gas

Objectives

In this experiment you will determine the molar mass and density of carbon dioxide gas using Avogadro's Law and the ideal gas equation.

Introduction

One of the first chemists to look for a physical interpretation behind the behavior of matter was Robert Boyle (1627–1691). In 1662 he published an account of the relationship between the pressure and volume of a gas. This was based upon experiments done separately by his assistant Robert Hooke and by Richard Towneley and Henry Power two years previously. This relationship is now known as Boyle's Law and states that the product of the pressure and volume of a gas at constant temperature is a constant.

Another law, commonly referred to as Charles' Law, was actually developed by Joseph Gay-Lussac (1778–1850) and published in 1802. This law states that the volume of a gas is proportional to the absolute temperature. The notion of an absolute zero temperature had been suggested a hundred years previously by Guillaume Amontons (1663–1705).

Finally, a third relationship was developed by Amadeo Avogadro (1776–1856) based upon experiments by Gay-Lussac. In 1811 Avogadro proposed that equal volumes of all gases, at the same temperature and pressure, contain equal numbers of molecules.

$PV = nRT$

P = pressure
V = volume
n = number of moles
T = temperature

Be sure to use the correct units in your calculations. If the pressure is in atm, the volume is in liters, and the temperature is in K, then the value of R is 0.08206 atm L K^{-1} mol^{-1}.

These three laws are combined together in the ideal gas law: $PV=nRT$, where R is the universal gas constant. It is a law for ideal gases; an ideal gas is one in which the intermolecular forces of attraction are zero and the size of the actual molecules of gas is zero. Under high pressure and low temperature conditions, the behavior of real gases deviates from the law. Fortunately, most gases at normal temperatures and pressures behave nearly ideally.

If the volume of some known gas can be determined, then under the same conditions of pressure and temperature, some unknown gas will contain exactly the same number of molecules. The difference in the masses of the two gases will be directly related to the difference in the molar masses of the two gases.

In this experiment you will measure the volume of a flask of air. Using the ideal gas law, you will determine the mass of air in the flask. Then you will measure the mass of the same volume of carbon dioxide gas. Comparison of the two masses will allow you to determine the molar mass of carbon dioxide.

The gases used in this experiment will be contaminated by water vapor. The air in the laboratory will be humid—it will contain water vapor in addition to nitrogen, oxygen, argon, carbon dioxide, and small amounts of other gases. Furthermore, in the first experiment, the carbon dioxide will be collected by displacement of water and will certainly be saturated by water vapor. The presence of this water vapor must be accounted for in the calculations you perform. This is most easily done by subtracting the partial pressure of water (the vapor pressure of water) from the total atmospheric pressure to determine the pressure due to the remaining gas. The partial pressure of water varies as the temperature changes and a table of the partial pressure of water *vs.* temperature is included at the end of this experiment.

The gas used as the standard in this experiment is ordinary air. Dry air has an average molar mass of 28.960 g mol^{-1}—a value, as expected, between the molar masses of nitrogen, 28, and oxygen, 32. Determining the masses of gases requires careful measurements—they're not very dense and large volumes have relatively little mass. However, with care, you should be able to determine the molar mass of carbon dioxide to within 1–2 percent.

Two experimental methods will be used to fill a flask with carbon dioxide; both should yield the same result. The first method involves the displacement of water from an inverted Erlenmeyer flask; carbon dioxide gas is less dense than water. In the second method carbon dioxide will displace air from an upright Erlenmeyer flask—air is less dense than carbon dioxide. In both methods care must be taken to obtain weighings as accurately as possible. Dry ice will be used as a convenient source of carbon dioxide gas.

Procedure

Work in pairs but prepare and submit individual data sheets. Write the name of your partner on your data sheet.

IMPORTANT!

Never use force to push stoppers into the necks of Erlenmeyer flasks. They are fragile and will break! The broken neck will cause severe cuts. Be gentle.

Experiment 1

1. Choose a clean 250 mL Erlenmeyer flask from your equipment drawer and a rubber stopper that fits securely. Use the same flask for all your experiments.

2. Weigh and record the mass of flask and stopper on an electronic balance.

3. Fill the flask to the brim with distilled water and insert the stopper gently but securely so that no air is trapped. Dry the outside of the flask thoroughly with a paper towel. Weigh the flask again.

4. The difference in the two masses is the mass of water. Since the density of water is 1.00 g/mL, the volume of the flask can now be calculated. You may wish to refill the flask with water to check the volume. There should be no need to reweigh the flask dry and empty—the first weighing on an electronic balance should be sufficiently accurate.

5. The next few steps should be repeated three times, perhaps more often if your first results lack precision.

6. Fill a 1L plastic beaker half full with distilled water. Immerse the neck of the water-filled Erlenmeyer flask, inverted, under the surface of the water and remover the stopper. Allow the stopper to fall to the bottom of the beaker but make sure that the stopper is facing up so that it can be reinserted easily when the flask is *gently* pushed down on it (see Figure 1).

7. Now drop a small pellet of dry ice into the beaker. Be careful when handling the dry ice—it is cold and can cause frostbite. A pellet about 1 cm in length and 1 cm in

Figure 1. Gas Collection

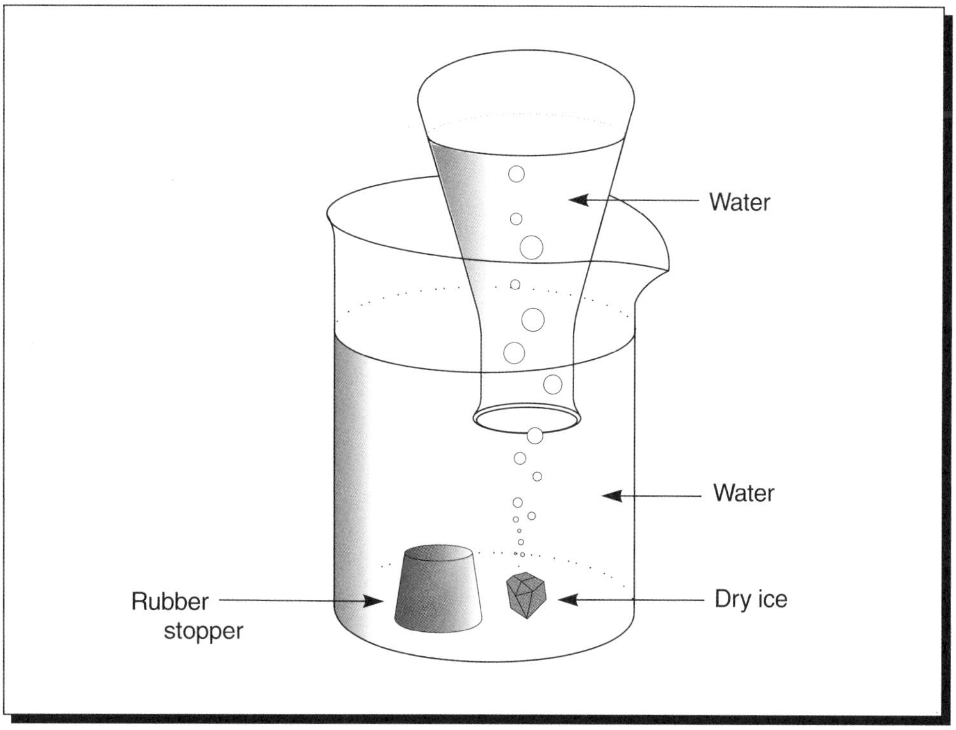

Water

Water

Rubber
stopper

Dry ice

IMPORTANT!

Never use force to push stoppers into the necks of Erlenmeyer flasks. They are fragile and *will* break! The broken neck will cause severe cuts. Be gentle.

Use the plastic beakers provided, not glass beakers!

Any water drops must remain in the flask! If you lose even the smallest drop, you should repeat the experiment. This experiment will test your ability to do careful work.

diameter is sufficient. Position the Erlenmeyer flask over the bubbles of carbon dioxide so that they enter the flask. When the flask is full of carbon dioxide, and the gas is escaping around the edge of the flask, push the flask firmly *but gently* onto the stopper under water. Very little, if any, water should be trapped in the flask.

8. Remove the stoppered flask and stand it upright. Any water drops on the inside should run to the bottom. Thoroughly dry the outside of the flask and the rubber stopper with paper towel. Very carefully and gently loosen the stopper and lift it slightly. Dry up any water around the rim and on the stopper with paper towel. Resecure the stopper, weigh the flask on an electronic balance and record the mass in Table 3 on your data sheet.

9. Some account must be taken of the small amount of water trapped in the flask. The best way to do this is to allow the carbon dioxide gas to escape and then reweigh the flask. With the flask in an upright position, remove the stopper. Then position the flask horizontally on your bench for 1 to 2 minutes. Replace the stopper and reweigh the flask. The difference between the two weighings will allow you to determine the mass of carbon dioxide originally in the flask.

10. Repeat the experiment twice for a total of three trials. Use a fresh beaker of water for each trial (it gets cold). The masses of carbon dioxide obtained in your three experiments should agree to within 25 mg. If they don't, you should consider a fourth trial. With a little practice you can achieve a precision within ±5-6 mg.

Experiment 2

1. This experiment serves as a check on the first method and requires only a single trial. Empty and thoroughly dry the same Erlenmeyer flask and stopper you used in the first experiment. Determine the mass of the flask and stopper. This mass should be the same as the mass you obtained at the beginning!

2. Place a small piece of dry ice (approximately 1 cm length by 1 cm diameter) in the Erlenmeyer flask and replace the stopper loosely to allow gas to escape. The dry ice will sublime slowly and fill the flask with carbon dioxide gas. Warm the flask in your hands and move the dry ice about in the flask. Shake the flask occasionally to break up the solid. It will take 15 minutes or so for the dry ice to completely sublime. While this is happening you may wish to start some of the calculations necessary in the analysis of your data.

3. When all but a few grains of solid carbon dioxide have sublimed, make sure that the outside of the flask is dry and place the flask (and the stopper) on an electronic balance. Watch the mass of the flask decrease to a constant value. Make a note of this mass on your data sheet.

To convert hPa to torr, multiply by 0.750.

4. Determine the ambient temperature in the laboratory to at least one decimal place. Measure the atmospheric pressure using a barometer. Look up the vapor pressure of water at the ambient room temperature; a table is provided at the end of this experiment.

Calculations

1. Determine the volume of the Erlenmeyer flask from the mass of water when the flask was filled.

The vapor pressures listed in Table 1 are the partial pressures of water vapor in air saturated with water at that particular temperature, i.e. 100% humidity. The air may not be saturated and 50% is perhaps a reasonable guess.

2. Determine the partial pressure of water in air from the vapor pressure table. If the ambient temperature is not listed, you should interpolate. If you know the relative humidity in the lab, then multiply the vapor pressure listed by this relative humidity. Otherwise assume a relative humidity of 50 percent and multiply by 0.50. Subtract this partial pressure of water from the total atmospheric pressure to obtain the pressure due to the air alone.

3. Using the ideal gas law, and knowing the pressure, the volume, the gas constant R, and the temperature, calculate the number of moles of air in the Erlenmeyer flask. If the average molar mass of air is 28.960, calculate the mass of air in the flask.

4. Calculate the apparent mass of carbon dioxide for each of the three runs. These values should be within 25 mg of each other. Calculate the average mass.

You may wonder why this correction wasn't made when the flask was weighed full of water. Should it have been? Answer the question on the reverse side on the data sheet.

5. However, when the flask is filled with carbon dioxide, there is no air present. So, to obtain the actual mass of carbon dioxide, the mass of air replaced by the carbon dioxide must be added to your result from step 4.

6. The number of moles n, in the ideal gas equation $PV = nRT$, equals the mass divided by the molar mass (m/M). So the molar mass $M = mRT/PV$. Use this relationship to calculate the molar mass of carbon dioxide from your data. The volume is the vol-

Be sure to use the correct units in your calculation. If the value of R is 0.08206 atm L K^{-1} mol^{-1}, then the pressure must be in atm, the volume in liters, and the temperature in K.

To convert torr to atm, divide by 760.

To convert mL to L, divide by 1000.

To convert °C to K, add 273.15.

ume of the flask as before. The pressure is the atmospheric pressure minus the vapor pressure of water at the ambient temperature. Since the carbon dioxide is saturated with water vapor, assume 100 percent humidity. The mass m is the actual mass of carbon dioxide.

7. Repeat the calculation using the mass of carbon dioxide from Experiment 2. Remember, as before, to add the mass of the displaced air but this time it is probably more reasonable to use a humidity of 50 percent again in correcting the atmospheric pressure.

8. From your results calculate the density of air and of carbon dioxide gas at room temperature and atmospheric pressure.

9. Answer the questions on your data sheet.

Table 1. The vapor pressure of water at various temperatures

Temp. °C	V.P. (Torr)	Temp. °C	V.P. (Torr)
15	12.8	23	21.1
16	13.6	24	22.4
17	14.5	25	23.8
18	15.5	26	25.2
19	16.5	27	26.7
20	17.5	28	28.3
21	18.6	29	30.0
22	19.8	30	31.8

$P = (988.5 \ hpa \ * \ .750)/760 = 0.9755$

123.645

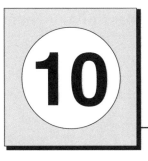

Determination of the Atomic Mass of Aluminum

Objectives

The objective of this experiment is to determine the atomic mass of aluminum from the stoichiometry of the reaction between aluminum and hydrochloric acid.

Introduction

Before the invention of the mass spectrograph by Francis William Aston in 1919, the atomic mass of an element was determined by stoichiometry. Known quantities of various substances were allowed to react and the masses of the reactants used or left over and the masses of products formed were measured. The atomic masses of the elements were then deduced from the data obtained.

The concepts of atoms and atomic "weight" were first introduced in the atomic theory of John Dalton published in 1803. Dalton made three assertions which are the basis for our understanding of the atomic nature of matter.

The chemical elements are composed of very minute individual particles of matter called atoms that preserve their individuality in all chemical changes.

All atoms of the same element are identical, particularly in their mass. Different elements have atoms differing in mass. Each element is characterized by the mass of its atom.

Chemical combination occurs by the union of the atoms of the elements in simple numerical ratios.

Dalton didn't know about isotopes—their existence for nonradioactive elements wasn't discovered until 1912 by J. J. Thomson and his student Francis Aston.

As the quality of balances improved and the understanding of how elements combine together improved, so did the accuracy of atomic masses. In 1826, a Swedish chemist, Jöns Jakob Berzelius, after more than ten years of work, published a remarkably accurate list of the atomic masses for fifty elements. While a number of compilations of atomic weights were published during this time, Berzelius' values are the most consistent with the values given today. He determined the atomic mass of aluminum to be 27.43 grams mol^{-1} in 1826. Today's value is 26.9815 grams mol^{-1}.

The purpose of this experiment is to determine the atomic mass of aluminum in much the same way as Berzelius did, with a gas measuring tube and a balance. This will be accomplished by studying the single replacement reaction that takes place when aluminum metal reacts with hydrochloric acid to produce hydrogen gas:

$$\text{metal} \quad + \quad \text{acid} \quad \rightarrow \quad \text{salt} \quad + \quad \text{hydrogen gas}$$

The gas generated will be collected by the downward displacement of water. This volume will be measured and then converted into moles of hydrogen. The atomic mass of aluminum will then be calculated from the mass of aluminum in the reaction and the number of moles of hydrogen collected.

Procedure

Work with a partner for this experiment but prepare and submit individual data sheets. Write the name of your partner on your data sheet.

Be careful with the glass tubing—it is fragile!

1. Tear a 4 to 5 inch length of aluminum foil from the roll of aluminum foil. Using scissors, cut the foil into three approximately 8 cm² pieces (2 cm x 4 cm).

2. Determine the mass of a piece of the aluminum foil on an electronic balance. The mass of the aluminum must not exceed 0.042 gram.

3. Fill a 1L plastic beaker with water leaving 1-2 cm at the top. Fill a 60 mL plastic syringe (with the tip cap on) with water and transfer to the beaker. The syringe should be completely full of water with the open end immersed in the water in the beaker as shown in Figure 1. Do not place the J-tube into the water at this time. You should figure out a way to get the syringe full of water into the beaker. The syringe must be full of water whether or not it is totally immersed in the beaker. Make a note of the temperature of the water on your data sheet.

4. Roll the aluminum foil into a loose coil and place it in a test tube. Add 3.0 ml of 4 M HCl. Press the stopper assembly firmly into the test tube with a twist to seal the system. Place the tip of the J-tube into the bottom of the syringe as shown in Figure 1. Check that all of the connections are tight to avoid a loss of hydrogen gas.

5. Swirl the test tube containing the acid and aluminum metal until the reaction begins to proceed rapidly. This may take three to four minutes depending on the temperature of the acid. Cool the reaction by placing the test tube into the water in the beaker. Continue to swirl the test tube while it is in the water to cool the mixture and slow the reaction. Wash any residual specks of unreacted aluminum metal from the sides of the test tube by tilting the test tube slightly and then shaking the solution until it contacts the metal. Record in Table 3 your observations of what happens to the contents of the test tube.

Figure 1. Apparatus for collecting hydrogen gas

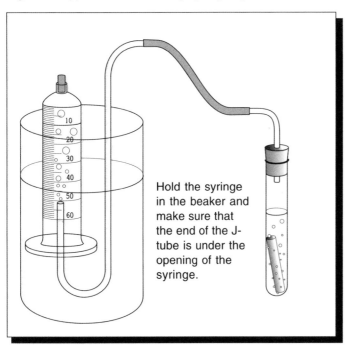

Hold the syringe in the beaker and make sure that the end of the J-tube is under the opening of the syringe.

6. When all the aluminum has reacted, you should equalize the pressure inside the syringe and the atmospheric pressure outside. With the tip of J-tube still inside the syringe, move both the syringe and the J-tube together until the open tip of the J-tube is just above the level of the water in the beaker. Now you can remove the J-tube from the syringe. When recording the volume of the gas in the syringe, always move the syringe up or down until the level of the water inside is nearly the same as the level of the water in the beaker. Record the level of water in the syringe to the nearest 0.5 mL. Remember that the syringe is upside down—the numbers increase downward. Record both the water temperature and the atmospheric pressure.

Figure 2. Equalizing the pressures inside and out

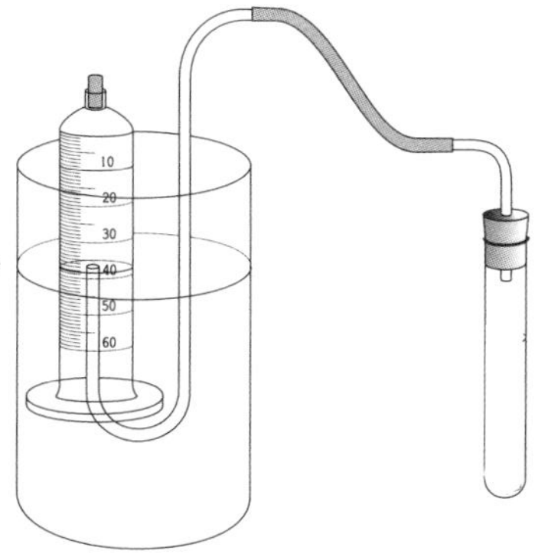

The level of water in the syringe and in the beaker should be the same. The tip of the J-tube should be just above the surface. Then remove the J-tube, equalize the levels and note the level of the water in the syringe.

You will be graded on both your accuracy and your precision.

7. Repeat the experiment at least twice and enter your data in Table 2.

8. Clean up the glass tubing and return it to your instructor.

Calculations

The barometer in your laboratory is calibrated in hPa (hectopascals).

To convert hPa to torr: multiply by 0.750.

1. The gas collected in the syringe is a mixture of hydrogen gas and water vapor. To obtain the partial pressure of hydrogen collected, the partial pressure of water vapor must be subtracted from the total (atmospheric) pressure. Look up in Table 1 the partial pressure of water (the vapor pressure) at the temperature of the water and enter in Table 2 on your data sheet. Subtract this from the atmospheric pressure to obtain the partial pressure of hydrogen in torr.

2. Convert the pressure in torr to atm.

3. The volume of hydrogen gas collected is determined from the reading of the water level in the syringe. This water level must be corrected because of the dead volume, uncalibrated, at the end of the syringe. This equals 1.20 mL and must be added to the water level reading to get the volume of hydrogen.

4. Now determine the number of moles of dry hydrogen gas generated by the reaction between the aluminum metal and the hydrochloric acid. This is accomplished using the ideal gas equation

$$PV = nRT$$

where P = partial pressure of the hydrogen gas in atmospheres
V = volume of H_2 collected in the syringe in liters
n = number of moles of hydrogen in the gas measuring tube
R = ideal gas constant, 0.08206 L atm K^{-1} mol^{-1}
T = the temperature of the gas in Kelvin

5. In Table 3 on your data sheet, write an equation for the reaction between aluminum metal and hydrochloric acid to produce hydrogen gas.

6. In Table 4, enter the number of moles of hydrogen gas produced (from Table 2) and calculate from the stoichiometry of the equation the number of moles of aluminum that must have been used in each experiment.

7. From the mass of aluminum used, and the number of moles calculated, determine the molar mass (the atomic mass) of aluminum.

8. Calculate the percent error in your result. The established atomic mass of aluminum is 26.98154. Divide the difference between your result and the established value by the established value and multiply by 100. This is the percent error.

Table 1. The vapor pressure of water at various temperatures

Temp. °C	V.P. (Torr)	Temp. °C	V.P. (Torr)
15	12.8	23	21.1
16	13.6	24	22.4
17	14.5	25	23.8
18	15.5	26	25.2
19	16.5	27	26.7
20	17.5	28	28.3
21	18.6	29	30.0
22	19.8	30	31.8

Red Cabbage, an Indicator for Acids and Bases

Objectives

In this experiment you will extract a dye from red cabbage and use this dye as an indicator to test the acidic or basic properties of some common household products. You will also prepare a sodium hydroxide solution for use in next week's experiment.

Introduction

There are many naturally occurring substances that change color in the presence of an acid or a base. Many have been known for a long time. For example, alizarin is an orange dye extracted from the root of the madder plant—it was used to dye wool in ancient Persia. Alizarin changes color from yellow in acids to red in almost neutral solutions. Litmus is a blue coloring matter that is extracted from lichens—its name means dye-moss. The blue color turns red in the presence of an acid and is restored to the blue color by alkali or base. This test has been so prevalent that the phrase "litmus test" is used to mean any definitive test for a particular criterion. Cochineal is yellow in acidic solutions but deep violet in basic solution. It is made from dried female cochineal insects.

5,000 insects are needed to make one ounce of the dry indicator.

Phenolphthalein is an indicator you may have already encountered—it is commonly used in acid-base titrations. This indicator is colorless in acidic solution but turns pink in basic solution.

The indicator that you will use in this experiment is derived from red cabbage (*brassica oleracea*). The cabbage contains a purple dye which is a mixture of 15 different anthocyanin molecules. These molecules change color in the presence of acid or base.

The name *anthocyanin* is derived from two Greek words, meaning *plant* and *blue*.

Anthocyanins are responsible for many of the fall foliage colors and are present in many plants. These are the molecules responsible for the red color of strawberries, poppies, and red cabbage, and the blue color of cornflowers and blueberries. Different plants contain varying amounts; dark blue pansies contain sixteen times as much anthocyanins as the purple delphinium. Anthocyanins are benzopyrylium salts (the ending -ium indicates the positive charge on the molecule) and three are illustrated in Figure 1. You can see the similarity in the three structures.

The anthocyanins, like most acid-base indicators, are fairly complex molecules that are weak acids or bases. The molecules exhibit one color when a hydrogen ion is attached to the molecule and a different color when the hydrogen ion is removed from the molecule. Anthocyanins in red cabbage change in color from red in acidic solution, to purple and green in slightly basic solution and then to yellow in strongly basic solution. The extract from red cabbage is an ideal indicator to study a range of different acidities.

The acidity of a substance is determined by its ability to produce hydronium H_3O^+ ions in solution. On the other hand, a basicity of a substance is determined by its ability to increase the number of hydroxide OH^- ions in solution. A solution that is

Figure 1. Anthocyanin molecules

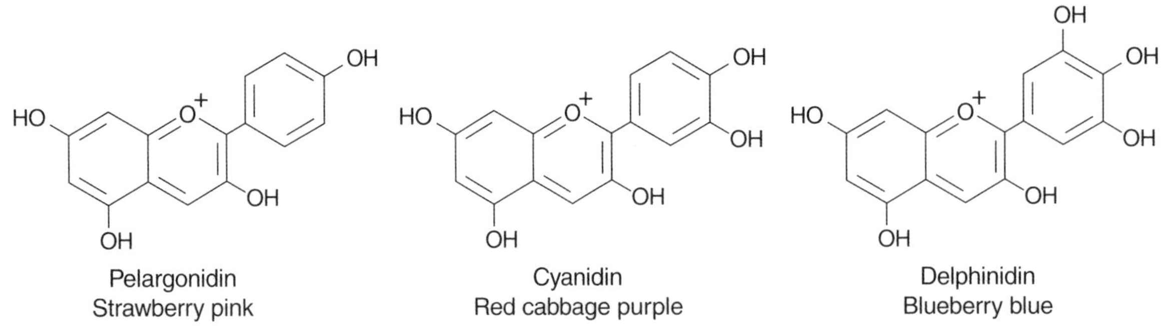

Pelargonidin
Strawberry pink

Cyanidin
Red cabbage purple

Delphinidin
Blueberry blue

The formula for the hydronium ion H_3O^+ is a way of emphasizing the fact that hydrogen ions H^+ do not exist by themselves in water solution; they are always attached to a water molecule. The hydronium ion itself is always strongly hydrogen bonded to surrounding water molecules. The number of water molecules associated with a single proton is large.

acidic contains more hydronium ions H_3O^+ than hydroxide ions OH^-. Similarly, a solution that is basic contains more hydroxide ions OH^- than hydronium ions H_3O^+. A neutral solution contains equal concentrations of the two ions. So the acidity or basicity of a solution is often expressed in terms of its hydronium ion concentration $[H_3O^+]$.

Since the value of the hydronium ion concentration can be very high or very low, it is convenient to take the logarithm of the concentration. This is called the pH of the solution. The pH is defined as $-\log_{10}[H_3O^+]$. A solution containing a concentration of H_3O^+ equal to 1.0×10^{-5}, for example, would have a pH of 5.0. So solutions with a *low* pH are acidic and solutions with a *high* pH are basic. The dividing line at room temperature is 7.0.

In pure water at room temperature, some high energy water molecules break apart as H_3O^+ and OH^- instead of just breaking up into two water molecules:

$$2\,H_2O \;\rightleftharpoons\; H_3O^+ \;+\; OH^-$$

The concentration of these two ions is very low, 10^{-7} M each, and the equilibrium constant K_w for the process is therefore 10^{-14}. Not many ions are formed. The pH of the pure water is defined as $-\log_{10}[H_3O^+]$, and therefore the pH of pure water at room temperature is 7.0—hence the dividing line between acid and base. If the temperature is raised, the concentrations of both ions increase (because more energy is available to break the bonds in the water molecules) and the pH decreases—but note that the water is still neutral! A neutral solution at body temperature has a pH a little below 7.0.

The product of the concentration of hydronium ion $[H_3O^+]$ and the concentration of hydroxide ion $[OH^-]$ is always the same at constant temperature (10^{-14} at room temperature). This means that as the hydronium ion concentration increases, the hydroxide ion concentration must decrease, and *vice versa*. Since the product of the two concentrations is always the same (at the same temperature), if you know one concentration you can always calculate the other.

Procedure

Work with a partner for this experiment but prepare and submit individual data sheets. Write the name of your partner on your data sheet.

A buffer solution is a solution that resists a change in pH when an acid or a base is added to it.

1. Your instructor will extract the anthocyanin dye from a red cabbage. This can be done by pulverizing the cabbage with distilled water in a food blender or by boiling a cabbage, cut into small pieces, in a 50:50 mixture of ethanol and water. The longer the pulverization or the boiling, the better the extraction will be. After straining the mixture, the anthocyanin solution will be ready for use.

2. Thirteen buffer solutions, each with a different pH from 1 to 13, are provided and each lab bench should make a reference pH set in a row of thirteen test tubes. Divide up the thirteen solutions and decide who should make up each reference color. Don't forget to place the tubes in the correct order—it would be a good idea to label them. For each solution, place 4 mL of the standard pH buffer solution in a test tube and add one full dropper of the anthocyanin indicator. It's a good idea to fill one test tube with 4 mL of water and mark the level with a Sharpie pen. Then fill the other test tubes to the same level.

3. Once the reference set is complete, examine the thirteen colors and record in Table 1 each color as descriptively as possible. Use colors or terms that you think most accurately describe each solution.

4. Eight clear dilute solutions of various reagents (in addition to distilled water) are supplied:

   - Distilled (pure) water
   - 0.5 M sodium chloride solution
   - 0.5 M sodium hydroxide solution
   - 0.5 M acetic acid solution
   - 0.5 M hydrochloric acid solution
   - 0.5 M sodium bicarbonate solution
   - 0.5 M potassium nitrate solution
   - 0.5 M ammonia solution
   - 0.5 M phosphoric acid solution

   Place approximately 4 mL of each solution in a test tube and test each solution with the red cabbage indicator. Again use a full dropper of the indicator solution for each reagent. Summarize your results in Table 2 on your data sheet.

5. Based upon your observations, make a rational classification of each of the solutions as strongly acidic, weakly acidic, neutral, weakly basic, or strongly basic. Compare the colors against your reference solutions and make a note of the pH you estimate for each solution.

6. Several household reagents are provided for testing. These might include an ammonia based window cleaner, vinegar, baking soda, lemon juice, a carbonated drink, detergent, antacid, bleach, milk of magnesia, salt, sugar, washing soda, bathroom cleaner, and/or several others. You can also bring to the lab any other

household reagents you wish to test. Test each product with the anthocyanin indicator, describe the color, and estimate the pH of the solution. Summarize the results in Table 3.

7. Answer the questions in Table 4.

Preparation of the sodium hydroxide solution

This week each student will prepare a sodium hydroxide solution required for next week's experiment. Next week's experiment is done individually and each student will need their own sodium hydroxide solution. You may start this procedure while completing Tables 3 and 4.

1. Boil 400 mL of distilled water for five minutes to remove any dissolved carbon dioxide (CO_2) gas. Cover the Erlenmeyer flask with a small watch glass after boiling the water or carbon dioxide from the air will redissolve. Cool to room temperature.

2. After the water has cooled, weigh approximately four grams of sodium hydroxide in your smallest beaker and add immediately to the cool water. Swirl the flask to dissolve the sodium hydroxide. The solution will be approximately 0.25 M.

3. As soon as the sodium hydroxide has dissolved and the solution is well mixed, transfer the solution to a 500 mL bottle. Screw on the cap part of the way and tighten it when the solution is cold.

4.083

Sodium hydroxide is caustic. Do not touch it! Do not leave any on the lab bench or on or around the balance!

Clean up any spills immediately!

The Determination of the Concentration of Acetic Acid in Vinegar

Objectives

Vinegar is a solution of acetic acid in water. In this experiment you will determine how much acetic acid is present in vinegar by titration of the vinegar against sodium hydroxide. First you will standardize the sodium hydroxide solution prepared last week by titration against hydrochloric acid. Then you will use the sodium hydroxide solution to determine the concentration of acetic acid in vinegar.

Introduction

Acid-base reactions in aqueous solution represent one of the most important types of chemical reaction. However, there are several different ways to define what an acid or a base is. One early and successful set of definitions was proposed by the Swedish chemist Svante Arrhenius in 1887.

Arrhenius defined an acid as a substance that produces hydrogen ions (H^+) in aqueous solution. The hydrogen ion is now more usually represented in solution as the hydronium ion H_3O^+. He defined a base as a substance that produces hydroxide ions (OH^-) in aqueous solution. According to the Arrhenius definitions then, an acid‐base reaction in aqueous solution is nothing more than the combination of hydronium ions and hydroxide ions to produce water molecules.

$$H_3O^+ \text{ (hydronium ion)} + OH^- \text{ (hydroxide ion)} \rightarrow 2H_2O$$

A broader and more useful set of definitions was proposed independently by Brønsted and Lowry in 1932. Like Arrhenius, they defined an acid as a hydrogen ion donor. However, they defined a base as a substance that accepts a hydrogen ion from the acid. Note that the hydroxide ion OH^- which accepts a hydrogen ion to form water in a neutralization reaction is an example of a Brønsted-Lowry base.

Two examples of acids are acetic acid and hydrochloric acid. However, these two acids differ remarkably in their behavior when dissolved in water. Hydrochloric acid is a strong electrolyte. This means that virtually all hydrochloric acid molecules are broken up into H^+ and Cl^- ions during the solution process:

$$HCl + H_2O \rightleftharpoons H_3O^+ + Cl^-$$
hydrochloric acid + water   hydronium ion + chloride ion

Acetic acid, on the other hand, is a weak electrolyte. Very few of the acetic acid molecules break up into ions when acetic acid is dissolved in water.

$$CH_3CO_2H + H_2O \rightleftharpoons H_3O^+ + CH_3CO_2^-$$
acetic acid + water   hydronium ion + acetate ion

The reason for this difference in behavior is a difference in the strength of the bond between the hydrogen and the rest of the molecule in the two acids. In each reaction there is a competition for the hydrogen ion. In the ionization of hydrochloric acid, water is a stronger base than the chloride ion Cl⁻ and the water therefore gets the hydrogen ion. In the ionization of acetic acid, on the other hand, the acetate ion $CH_3CO_2^-$ is a stronger base than water and it keeps the hydrogen ion.

A 0.10 M solution of a strong acid like hydrochloric acid will contain a greater concentration of hydronium ions than a 0.10 M solution of acetic acid. The pH of the strong acid will therefore be lower than the pH of a weak acid of equivalent concentration.

The reaction between an acid and a base is called *neutralization* because the solution which results no longer contains acid or base but only the salt. But very often the solution that results from the reaction of equivalent amounts of acid and base is not neutral. If the base is a strong base like sodium hydroxide and the acid is a weak acid like acetic acid, then the *equivalence point* occurs when the solution is basic.

$$CH_3CO_2H + NaOH \rightarrow NaCH_3CO_2 + H_2O$$
weak acid + strong base → salt + water

The salt produced in this reaction is sodium acetate and the acetate ion hydrolyses (reacts with water) to produce the weak electrolyte acetic acid and hydroxide ions. It is these hydroxide ions that make the solution basic.

The graphs in Figures 1 and 2 (on the next page) are called *titration curves*. The lines follow the change in the pH of the solution during the neutralization process. The experimental procedure is called titration. Notice the higher pH at the starting point and the basic equivalence point when the acid is weak.

$pH = -\log[H^+]$

The equivalence point in a titration is detected using an *indicator*. An indicator is a substance that has different colors in acidic solution and basic solution. The color changes because the structure of the molecule changes as the pH changes.

The equivalence point in a titration is that point at which equivalent amounts of acid and base have been combined.

There are a variety of these substances available, each changing color over different pH ranges, so that it is easy to pick an appropriate one for a particular titration. These substances provide a visual and very precise indication of the equivalence point in a titration because, as shown in Figures 1 and 2, the pH changes rapidly at the equivalence point.

The color change in a titration is referred to as the *end-point*. If the indicator is a good one, the end point will coincide with the equivalence point. It is possible of course to choose an indicator that changes color at a pH that does not correspond to the equivalence point. In this case the end-point and equivalence point will not coincide. In this experiment you will use an indicator called phenolphthalein. The phenolphthalein indicator changes color from colorless to pink-red over the pH range 7 to 9.

You will first standardize your solution of sodium hydroxide by titration against a hydrochloric acid solution of known concentration.

The second part of the experiment will be the titration of this standardized sodium hydroxide solution against a known quantity of vinegar to find the concentration of acetic acid in vinegar.

Figure 1. pH change during a strong acid–strong base titration

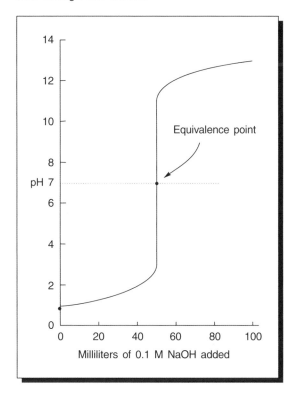

Figure 2. pH change during a weak acid–strong base titration

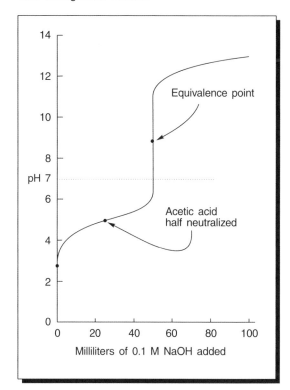

Procedure

Work on your own for this experiment.

Standardization of the sodium hydroxide solution

1. Gently boil about 400 mL of distilled water in an Erlenmeyer flask for about ten minutes to remove any dissolved carbon dioxide. Carbon dioxide reacts with sodium hydroxide to produce sodium carbonate. This interferes with the titration. Allow the water to cool with a watch glass on top of the flask to keep carbon dioxide out.

2. Wash a 250 mL Erlenmeyer flask thoroughly and rinse it with distilled water. You do not have to dry the inside of the flask.

3. From the autopipetter, pipet 10 mL of the standard HCl solution into the flask. Your instructor will show you how to use the autopipetter. Add about 20 to 30 mL of the boiled distilled water to the flask.

4. Select a buret and clean it thoroughly. Rinse with tap water and then rinse further with five portions of distilled water. Water must run freely from the tip and should flow in a continuous film from the inside of the buret without leaving any droplets adhering to the glass. If it does not, clean it again. Make sure the buret does not leak and that the stopcock turns freely.

5. Rinse the buret with three small (no more than 3 mL) portions of your sodium hydroxide solution. Then fill the buret to near the top of the graduation marks. Run out a small amount to flush any air bubbles from the buret tip. Discard

To make it easier to read the meniscus level, prepare a small white card or piece of paper. Hold the card behind the buret. The curve of the meniscus will appear black against the white card.

If you're left-handed, you may prefer to swirl the flask with your left hand.

If you pass the equivalence point, you must repeat the titration.

this; do not put it back into the stock solution! Remember to keep the Erlenmeyer flask or beaker containing sodium hydroxide solution covered.

6. Read and record on the *third* line of Table 1 the initial level of liquid in the buret to two decimal places (0.01 milliliter). Determine the level at the *bottom* of the meniscus and estimate the last decimal place as precisely as possible. Avoid parallax errors by keeping your eye level with the meniscus.

7. Add 3–4 drops of phenolphthalein indicator to the Erlenmeyer flask containing hydrochloric acid solution. Slowly add sodium hydroxide solution from the buret to the Erlenmeyer flask while gently swirling its contents as shown in Figure 3. As the sodium hydroxide solution is added, a pink color appears where the drops enter the solution. As the equivalence point is approached, this pink color persists for a longer and longer time. Near the equivalence point the inside of the Erlenmeyer flask should be rinsed down with distilled water and the sodium hydroxide solution should be added drop by drop. The equivalence point is reached when one drop turns the solution from colorless to a faint but persistent pink.

8. Allow the buret level to stabilize for 30 seconds and then read and record the level as before to two decimal places. Calculate the volume of sodium hydroxide used and enter the result in Table 1.

9. Repeat this procedure two times for a total of three titrations. You will be graded on the precision of your data.

Figure 3. Titration

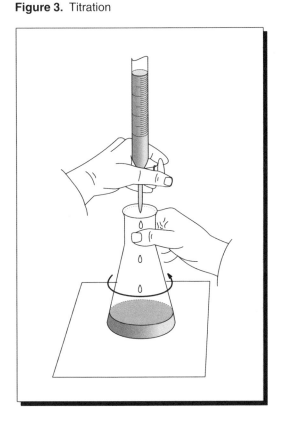

Titration of vinegar

11. Clean the three Erlenmeyer titration flasks thoroughly and rinse with distilled water three or four times. There is no need to dry the insides of the flasks.

12. Pipet into each of the flasks 5 mL of vinegar using an autopipetter as before. Add 20 to 30 mL of the boiled distilled water to each flask.

13. Refill your buret with sodium hydroxide solution and read and record the initial level as before.

14. Add phenolphthalein indicator and proceed with three titrations. Again, swirl the flask continuously during the titration, allow the buret level to stabilize before taking a reading and remember that all readings must be made to two decimal places. The three titrations must agree to within 1 percent.

15. Calculate the molarity of the vinegar. Details of this calculation are provided in the calculation section. Enter the result on your data sheet in Table 2.

Calculations

The calculations in this experiment are based on the mole concept. One mole of sodium hydroxide will produce one mole of hydroxide ions in solution. One mole of either hydrochloric acid or acetic acid contains one mole of hydrogen ions.

One mole of either acetic acid or hydrochloric acid will be neutralized by one mole of sodium hydroxide in a titration.

The concentration of the acid or base solution can be expressed in moles per liter or in millimoles per milliliter:

    Molarity = moles/liter = mmoles/milliliter

If the molarity and volume are known, then the number of moles or millimoles can be calculated:

    Moles = molarity × volume in liters

    Millimoles = molarity × volume in milliliters

In all the titrations in this experiment, the number of moles of base required in the titration equals the number of moles of acid used.

Calculation of the molarity of the sodium hydroxide solution

At the equivalence point, the number of moles of hydrochloric acid (HCl) equals the number of moles of sodium hydroxide (NaOH):

Molarity of HCl × volume of HCl = molarity of NaOH × volume of NaOH
     known   10 mL   unknown   buret reading in mL

Calculation of the molarity of the acetic acid (vinegar) solution

At the equivalence point, the number of moles of acetic acid in the vinegar equals the number of moles of NaOH:

Molarity of vinegar × volume of vinegar = molarity of NaOH × volume of NaOH
     unknown   5 mL   now known   buret reading in mL

A | | | | | | | |

NAME

GRADING		
Precision of standardization of NaOH	/2
Precision in titration of vinegar	/3
Accuracy of result	/3
Titration technique	/2
Total	/10

(0)
(1)
(2)
(3)
(4)
(5)
(6)
(7)
(8)
(9)
(10)

Table 1. Standardization of the NaOH solution

	Trial 1	Trial 2	Trial 3	Average
Molarity of the HCl solution				
Final buret reading (mL)				
Initial buret reading (mL)				
Volume of NaOH required (mL)				
Molarity of the NaOH solution				

Table 2. Titration of vinegar with standard NaOH solution

	Trial 1	Trial 2	Trial 3	Average
Final buret reading (mL)				
Initial buret reading (mL)				
Volume of NaOH required (mL)				
Molarity of acetic acid in vinegar				

CONCENTRATION OF ACETIC ACID IN VINEGAR

A Cycle of Copper Reactions

13

Objectives

In this experiment you will perform a series of reactions involving copper and its compounds. You will investigate the ways in which one copper compound can be converted into another. The series of reactions starts with copper metal and ends with copper metal. One of the objects of the experiment is to finish with the same quantity of copper with which you started.

Introduction

Ira Remsen, the first professor of chemistry at Johns Hopkins University, became interested in chemistry when he read in an inorganic chemistry textbook that "nitric acid acts upon copper." Intrigued by the phrase "acts upon," he determined to find out for himself what it meant. He took a copper penny, placed it on the desk in the doctor's office where he was working at the time, and dropped a few drops of concentrated nitric acid onto it. A green-blue liquid foamed and fumed over the cent; the air above the reaction became colored dark red. The vigorous reaction continued, not only between the nitric acid and the penny, but also between the nitric acid and the desk, and then between nitric acid and the fingers of Ira Remsen when he attempted to pick up the penny to throw it out the window. Thus Ira Remsen learned the oxidizing power of nitric acid.

Figure 1. A cycle of copper reactions

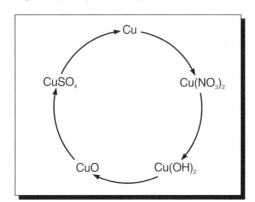

You will repeat this experiment—but without using your fingers—as the first of a series of steps involving the chemistry of copper. In the experiment you will start with copper metal and then progress through a series of reactions that change the copper through a series of its compounds. Eventually you will finish with copper metal. If your technique is good, you will end up with the same amount of copper as that with which you started.

The series of reactions is illustrated in Figure 1. Each step in this cycle is driven to completion because either a gas is evolved, or a precipitate is formed, or some other process drives the reaction forward.

The first reaction is the oxidation of copper metal to copper(II) using nitric acid. The copper(II) forms the soluble salt copper(II) nitrate, $Cu(NO_3)_2$, and you will see its characteristic blue color. Nitric acid is reduced to form nitrogen dioxide. An equation that represents the reaction is:

$$Cu(s) \ + \ 4HNO_3(aq) \ \rightarrow \ Cu(NO_3)_2(aq) \ + \ 2NO_2(g) \ + \ 2H_2O(l)$$

The second step is the precipitation of copper(II) hydroxide from the solution by addition of sodium hydroxide. Unlike the nitrate salt, the hydroxide of copper(II) is insoluble:

$$Cu(NO_3)_2 (aq) \ + \ 2NaOH (aq) \ \rightarrow \ Cu(OH)_2 (s) \ + \ 2NaNO_3 (aq)$$

Heating the hydroxide causes loss of water and converts the hydroxide into copper(II) oxide:

$$Cu(OH)_2 (s) \ + \ Heat \ \rightarrow \ CuO (s) \ + \ H_2O (g)$$

Copper(II) oxide is then treated with sulfuric acid to produce the soluble salt copper(II) sulfate, $CuSO_4$. Its solution again has the characteristic color of copper(II).

$$CuO (s) \ + \ H_2SO_4 (aq) \ \rightarrow \ CuSO_4 (aq) \ + \ H_2O (l)$$

Finally, in the fifth step, the copper(lI) sulfate solution is treated with zinc metal. Another reduction-oxidation reaction occurs but this time copper(II) is reduced to copper metal and zinc metal is oxidized to zinc(II) ions:

$$CuSO_4 (aq) \ + \ Zn (s) \ \rightarrow \ ZnSO_4 (aq) \ + \ Cu (s)$$

Procedure

You must work on your own for this experiment.

Nitric acid will cause severe burns. The fumes of nitric oxide and nitrogen dioxide given off are extremely toxic (200 ppm can be FATAL!). You must perform this operation under the hood. You must make sure the hood is ON. You must wear your safety goggles. You must NOT breathe the vapor.

Bumping is a sudden and vigorous boiling. The contents of the beaker are often lost when bumping occurs.

1. Weigh approximately 0.50 gram of copper metal as precisely as possible. Place it in a 250 mL beaker.

2. Under the fume hood, very carefully add 4.0 to 5.0 mL of concentrated nitric acid to the beaker. Note that concentrated nitric acid is hazardous! If necessary, use a glass rod to push the copper under the liquid. If you use a glass rod, be sure not to remove it from the beaker without proper washing; otherwise product will be lost!

3. After the reaction has finished and all the copper metal has been oxidized, add approximately 100 mL of distilled water all at once.

4. Stir the solution with a glass rod and add 30 mL of 3.0 M sodium hydroxide solution to precipitate the copper(II) as copper(II) hydroxide, $Cu(OH)_2$.

5. Add a single boiling chip. Heat the beaker gently on a hot plate with constant stirring to convert the copper(II) hydroxide to copper(II) oxide, CuO. The mixture has a tendency to bump!

   If this happens, some copper oxide may be lost and you should start again. Stirring the mixture with a glass rod and adding a boiling chip will prevent bumping. Do not allow the solution to boil vigorously. You will notice the change in the appearance of the suspension in the beaker.

6. While this beaker is being heated on the hot plate, heat another beaker of distilled water to boiling.

7. Remove the beaker containing the copper(II) oxide from the hot plate and allow the copper(II) oxide to settle. Decant the clear supernatant liquid into another

Decant the liquid into another beaker just in case some copper oxide is accidentally lost—you can then recover it. Afterwards, the supernatant liquid can be poured down the sink.

Use the same technique any time you decant a supernatant liquid.

Concentrated hydrochloric acid is hazardous! Do not breathe the fumes from it. Do not get it on your skin.

Organic solvents such as methanol (CH_3OH) and acetone [($CH_3)_2CO$] should be disposed of into waste containers.

Acetone is very volatile and is easily evaporated.

Percent yield =

$$\frac{\text{Actual yield}}{\text{Theoretical yield}} \times 100$$

beaker. Add 200 mL of hot distilled water from the other beaker you have been heating. Allow the copper(II) oxide to settle again and decant the supernatant liquid again. Do not pour out any of your copper oxide!

8. Add 15 mL of 6.0 *M* sulfuric acid to convert the copper(II) oxide into copper(II) sulfate, $CuSO_4$.

9. Again place the beaker under the fume hood and add 2.5 grams of 30-mesh zinc metal all at once to the solution. Stir the solution until it becomes colorless. When the evolution of hydrogen has more or less finished, allow the solid product to settle. Decant the supernatant liquid.

10. Replace the beaker under the hood. Add 5 mL of distilled water and 10 mL of concentrated hydrochloric acid. This procedure removes any unused zinc metal.

11. When visible reaction has ceased, warm the solution without boiling it to ensure that the reaction is complete. When no further gas evolution is observed, remove the beaker from the hot plate. Wash the product with at least three 5 mL portions of distilled water. Decant the wash water after each washing.

12. Remove the boiling chip very carefully with forceps and wash it with distilled water over the beaker. Decant the water.

13. To dry the copper metal quickly, first wash it with 5 mL of methanol to displace the water. Allow the copper metal to settle and then decant the methanol. Next wash the metal with 5 mL of acetone to displace the methanol. Allow the copper to settle and decant the acetone. Do not use more than 5 mL of each.

14. Dry the copper metal in a beaker on a hot plate set at a low temperature. Remove the beaker from the hot plate, allow it to cool, transfer the product to a weighing paper, and weigh it. Transfer the copper metal back to the beaker, heat it for four or five more minutes, allow it to cool again, and then reweigh it to check for any further loss of mass. *You must show your copper product to your instructor.*

15. Calculate the percent yield to see how efficiently you performed these operations. If your yield is substantially less than 100 percent, estimate the stage at which you think you lost some copper.

References

Todd, D. and Hobey, W. D., *J. Chem. Educ.* **62**, 177 (1985)

Condike, G. F., *J. Chem. Educ.* **52**, 615 (1975)

A [ ][ ][ ][ ][ ][ ][ ][ ]

(A) (0)(0)(0)(0)(0)(0)(0)(0)
(1)(1)(1)(1)(1)(1)(1)(1)
(2)(2)(2)(2)(2)(2)(2)(2)
(3)(3)(3)(3)(3)(3)(3)(3)
(4)(4)(4)(4)(4)(4)(4)(4)
(5)(5)(5)(5)(5)(5)(5)(5)
(6)(6)(6)(6)(6)(6)(6)(6)
(7)(7)(7)(7)(7)(7)(7)(7)
(8)(8)(8)(8)(8)(8)(8)(8)
(9)(9)(9)(9)(9)(9)(9)(9)

NAME [ ]

GRADING	Completion and performance	/2
	Yield and appearance of product	/4
	Explanation of observations (Table 2)	/4
	Total	/10

(0)
(1)
(2)
(3)
(4)
(5)
(6)
(7)
(8)
(9)
(10)

Table 1. Weighing data

Mass of copper metal at the start	g
Mass of copper metal at the end	g
Percent copper recovered	%

Table 2. Description and explanation of your observations

Describe and explain your observations. Mention any stage at which copper may have been lost. Do not rewrite the equations from the introduction. Describe and explain what you see!

1. The reaction of copper with nitric acid

2. The precipitation of copper(II) hydroxide

3. The preparation of the copper(II) oxide

4. The preparation of copper(II) sulfate

5. The reduction of copper(II) with zinc